Any number can be written as a product of prime factors. This means the number is written it using only prime numbers multiplied together.

Lowest Common Multiple (LCM)

This is the lowest number that is a multiple of two numbers.

Example

Find the LCM of 60 and 96.

- Write the numbers as products of their prime factors.

  $60 = \boxed{2} \times \boxed{2} \times \boxed{3} \times 5$

  $96 = \boxed{2} \times \boxed{2} \times 2 \times 2 \times 2 \times \boxed{3}$

- 60 and 96 have a common prime factor of $2 \times 2 \times 3$, so it is only counted once.
- The LCM of 60 and 96 is

  $2 \times 2 \times 2 \times 2 \times 2 \times 3 \times 5$

  $= 480$

Progress check

1. Write these numbers as a product of prime factors:
   a) 50
   b) 360
   c) 16

2. Decide whether these statements are true or false.
   a) The HCF of 20 and 40 is 4.
   b) The LCM of 6 and 8 is 24.
   c) The HCF of 84 and 360 is 12.
   d) The LCM of 24 and 60 is 180.

3. Find the HCF and LCM of 36 and 48.

D0185672

FRACTIONS AND RECURRING DECIMALS

FOUR RULES OF FRACTIONS

Addition +

$\frac{5}{9} + \frac{1}{7}$

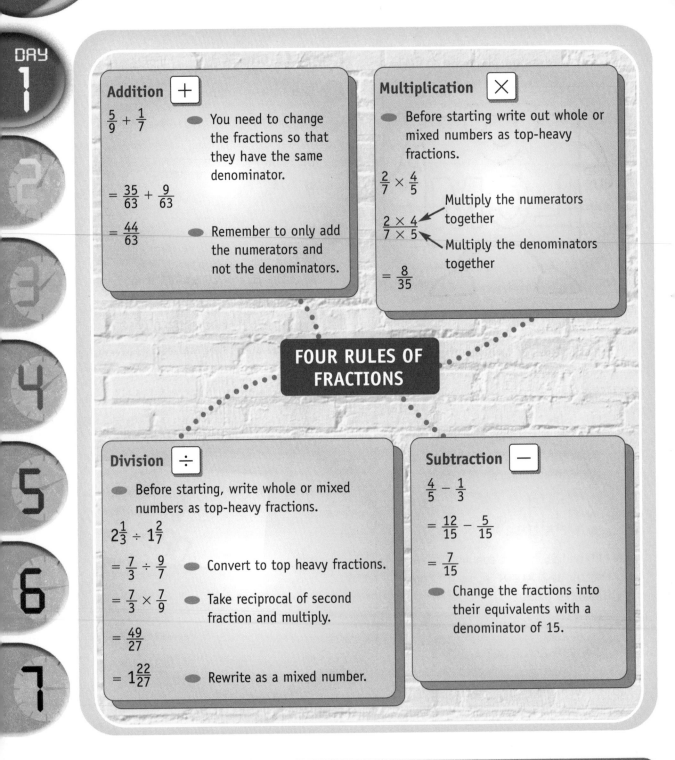

- You need to change the fractions so that they have the same denominator.

$= \frac{35}{63} + \frac{9}{63}$

$= \frac{44}{63}$

- Remember to only add the numerators and not the denominators.

Multiplication ×

- Before starting write out whole or mixed numbers as top-heavy fractions.

$\frac{2}{7} \times \frac{4}{5}$

$\frac{2 \times 4}{7 \times 5}$

Multiply the numerators together

Multiply the denominators together

$= \frac{8}{35}$

Division ÷

- Before starting, write whole or mixed numbers as top-heavy fractions.

$2\frac{1}{3} \div 1\frac{2}{7}$

$= \frac{7}{3} \div \frac{9}{7}$ — Convert to top heavy fractions.

$= \frac{7}{3} \times \frac{7}{9}$ — Take reciprocal of second fraction and multiply.

$= \frac{49}{27}$

$= 1\frac{22}{27}$ — Rewrite as a mixed number.

Subtraction −

$\frac{4}{5} - \frac{1}{3}$

$= \frac{12}{15} - \frac{5}{15}$

$= \frac{7}{15}$

- Change the fractions into their equivalents with a denominator of 15.

A fraction is part of a whole one. The top number is the numerator, and the bottom number is the denominator.

Changing recurring decimals to fractions

Recurring decimals are **rational** numbers, hence we can change them to fractions.

Examples

1. Change $0.\dot{1}\dot{5}$ to a fraction in its lowest terms.

Let $x = 0.151515 \ldots\ldots$ ①

then $100x = 15.151515 \ldots\ldots$ ②

Multiply by 10^n, where n is the length of the recurring pattern

② − ① $99x = 15$

$$x = \frac{15}{99}$$

$$x = \frac{5}{33} \quad (\text{check } 5 \div 33 = 0.\dot{1}\dot{5})$$

2. Change $0.3\dot{7}$ into a fraction

$x = 0.3777 \ldots\ldots$ ①

$10x = 3.7777 \ldots\ldots$ ②

Multiply by 10, so that the part that is not recurring is in the units position.

$100x = 37.7777 \ldots$ ③ multiply by 100

Subtract equation ② from equation ③.

③ − ② $90x = 34$

$$x = \frac{34}{90}$$

$$x = \frac{17}{45}$$

Always check to see if your fraction simplifies.

Progress check

1 Work out the following:

a) $\frac{2}{3} + \frac{1}{5}$

b) $\frac{6}{7} - \frac{1}{3}$

c) $\frac{2}{9} \times \frac{5}{7}$

d) $\frac{3}{11} \div \frac{22}{27}$

2 Change the following recurring decimals into fractions. Write each fraction in its simplest form.

a) $0.\dot{7}$

b) $0.\dot{2}\dot{1}\dot{5}$

c) $0.3\dot{5}$

REPEATED PERCENTAGE CHANGE AND COMPOUND INTEREST

DAY 1

Repeated percentage change

A car was bought for £12 500. Each year it depreciated in value by 15%. What was the car worth after three years?

You must remember not to do: 3 × 15% = 45% reduction over 3 years!

Method 1

- Find 100 − 15 = 85% of the value of the car first.

  Year 1 $\frac{85}{100} \times £12\,500 = £10\,625$

- Then work out the value year by year. (£10 625 depreciates in value by 15%.)

  Year 2 $\frac{85}{100} \times £10\,625 = £9031.25$

  (£9301.25 depreciates in value by 15%.)

  Year 3 $\frac{85}{100} \times £9031.25 = £7676.56$

Method 2

- A quick way to work this out is by using a multiplier.
- Finding 85% of the value of the car is the same as multiplying by 0.85.

  Year 1: 0.85 × £12 500

  Year 2: 0.85 × £10 625

  Year 3: 0.85 × £9031.25

- This is the same as working out $(0.85)^3 \times £12\,500 = £7676.56$

Compound interest is where the bank pays interest on the interest earned as well as on the original money.

Compound interest

Charlotte has £3200 in her savings account and compound interest is paid at 3.2% p.a. How much will she have in her account after four years?

100 + 3.2 = 103.2%

= 1.032 This is the multiplier.

Year 1: 1.032 × £3200 = £3302.40

Year 2: 1.032 × £3302.40 = £3408.08

Year 3: 1.032 × £3408.08 = £3517.14

Year 4: 1.032 × £3517.14 = £3629.68

Total = £3629.68

A quicker way is to multiply £3200 by $(1.032)^4$

number of years

$£3200 \times (1.032)^4 = £3629.68$

original   multiplier

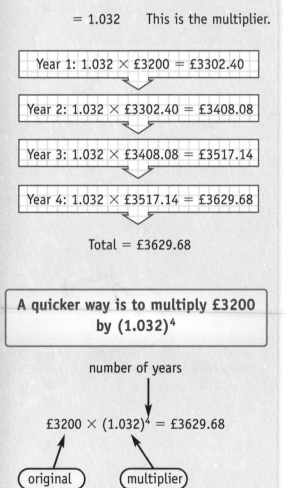

Progress check

1. Reece has £5200 in the bank. If compound interest is paid at 2% p.a., how much will he have in his account after 3 years?

2. Complete this statement:
Some money is invested for 2 years. If compound interest is paid at 2.7% p.a. the multiplier would be
....

3. Mr Singh bought a flat for £85 000 in 1999. The flat rose in value by 12% in 2000 and 28% in 2001. How much was the flat worth at the end of 2001?

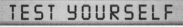

REVERSE PERCENTAGE PROBLEMS

The price of a television is reduced by 20% in the sales.
It now costs £350. What was the original price?

- The sale price is 100% − 20% = 80% of the pre-sale price (x)
- 80% = 0.8 This is the multiplier.
- $0.8 \times x = £350$

$$x = \frac{£350}{0.8}$$

Original price = £437.50

Check

original price —— $\times$ **0.8** ——→ **new price**

new price —— $\div$ **0.8** ——→ **original price**

Does the answer sound sensible?
Is the original price more than the sale price?

A telephone bill costs £169.20 including VAT at $17\frac{1}{2}$%.
What is the cost of the bill without the VAT?

- The telephone bill of £169.20 represents 100% + 17.5% = 117.5% of the original bill (x).
- 117.5% = 1.175 This is the multiplier.
- $1.175 \times x = £169.20$

$$x = \frac{£169.20}{1.175}$$

Original bill = £144

Check

original bill —— $\times$ **1.175** ——→ **new bill**

new bill —— $\div$ **1.175** ——→ **original bill**

These are when the original quantity is calculated. They are quite tricky so think carefully!

The price of a washing machine is reduced by 15% in the sales. It now costs £323. What was the original price?

- The sale price is 100% − 15% = 85% of the pre-sale price (x).
- 85% = 0.85   This is the multiplier.
- $0.85 \times x = £323$

$$x = \frac{£323}{0.85}$$

Original price = £380

Progress check

1 Each item listed below includes VAT at 17.5%. Work out the original price of the item.

a) a pair of shoes: £62 £51.15
b) a coat: £125 £103.13
c) a suit: £245 £202.13
d) a TV: £525 £433.13

2 In the winter sales the price of the items below are reduced by 15%, and the new prices are given. Joseph works out the original prices and writes them below. Decide whether Joseph is correct.

a) CD player
£60 Yep
original price:
£70.59

b) Mountain bike
£240
original price: No
£276

c) Sweater
£30
original price: Yep
£35.29

INDICES

15 MINS

DAY 1

$$a^b$$

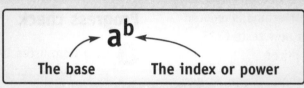

The base — The index or power

The **base** has to be the **same** when the laws of indices are applied. The laws of indices can be used for numbers or in algebra.

Laws of indices

$$a^n \times a^m = a^{n+m}$$

$$a^n \div a^m = a^{n-m}$$

$$(a^n)^m = a^{n \times m}$$

$$a^0 = 1$$

$$a^1 = a$$

$$a^{-1} = \frac{1}{a^1}$$

$$a^{\frac{1}{m}} = {}^m\sqrt{a}$$

$$a^{\frac{m}{n}} = ({}^n\sqrt{a})^m$$

Examples with numbers

1. Simplify the following, leaving your answer in index notation.

   a) $5^2 \times 5^3 = 5^{\mathbf{2+3}} = 5^5$

   b) $8^{-5} \times 8^{12} = 8^{\mathbf{-5+12}} = 8^7$

   c) $(2^3)^4 = 2^{\mathbf{3x4}} = 2^{12}$

2. Evaluate:

   Evaluate means to work out.

   a) $4^2 = 4 \times 4 = 16$

   b) $5^\circ = 1$

   c) $3^{-2} = \frac{1}{3^2} = \frac{1}{9}$

   d) $36^{\frac{1}{2}} = \sqrt{36} = \pm 6$

   e) $8^{\frac{2}{3}} = ({}^3\sqrt{8})^2 = 2^2 = 4$

Examples with algebra

1. Simplify the following:

a) $a^4 \times a^{-6} = a^{4-6} = a^{-2} = \frac{1}{a^2}$

b) $5y^2 \times 3y^6 = 15y^8$

The numbers are multiplied

The indices are added

c) $(4x^3)^2 = 16x^6$

Remember to square the 4 as well.

d) $(2x)^{-3} = \frac{1}{(2x)^3} = \frac{1}{8x^3}$

2. Simplify:

a) $\frac{15b^4 \times 3b^7}{5b^2} = \frac{45b^{11}}{5b^2} = 9b^9$

b) $\frac{16a^2b^4}{4ab^3} = 4ab$

Progress check

1 Simplify the following, leaving your answers in index form.

a) $6^3 \times 6^5$

b) $12^{10} \div 12^{-3}$

c) $7^{10} \div 7^{-14}$

d) $(5^2)^3$

e) $64^{\frac{5}{6}}$

2 Simplify the following:

a) $2b^4 \times 3b^6$

b) $8b^{-12} \div 4b^4$

c) $(3b^4)^2$

d) $\frac{9b^6 \times 2b^5}{3b^{-3}}$

e) $(5x^2y^3)^{-2}$

STANDARD INDEX FORM

When written in standard form a number will be written as $a \times 10^n$

A number between 1 and 10
$1 \leqslant a < 10$

The value of n is the number of places the digits have moved to return the number to its original value.

Example

Write 2 730 000 in standard form.

- Move the decimal point to between the 2 and 7, to give 2.73 ($1 \leqslant 2.73 < 10$).
- Count how many spaces the decimal point has to move to restore the original number.

2 7 3 0 0 0 0 (6 places)

So $2\,730\,000 = 2.73 \times 10^6$

- **If the number is big, n is positive.**
- **If the number is small, the n is negative.**

Example

Write 0.000046 in standard form.

4.6×10^{-5}

- Put the decimal point between the 4 and 6.
- Move the decimal point back five places to restore the original number.
- The value of n is negative.

Standard index form is used to write very large or very small numbers in a simpler way.

To put a number written in standard form into your calculator you use the EXP or EE key.

$(2 \times 10^3) \times (6 \times 10^7) = 1.2 \times 10^{11}$

This would be keyed in as

2 EXP 3 × 6 EXP 7 =

On a non-calculator paper you can use indices to help work it out.

$2 \times 10^3 \times 6 \times 10^7$

$= 2 \times 6 \times 10^3 \times 10^7$

$= 12 \times 10^{3+7}$

$= 12 \times 10^{10}$

$= 1.2 \times 10 \times 10^{10}$

$= 1.2 \times 10^{11}$

Progress check

1. Write in standard form:
   a) 64 000 6.4×10^4
   b) 271 000 2.71×10^5
   c) 0.00046 4.6×10^{-4}
   d) 0.000000074 7.4×10^{-8}

3. Without a calculator, work out the following. Leave in standard form.
   a) $(3 \times 10^4) \times (4 \times 10^6)$
   b) $(6 \times 10^{-5}) \div (3 \times 10^{-4})$
   c) $(5 \times 10^6) \times (7 \times 10^9)$

3. Work these out on a calculator:
   a) $(4.6 \times 10^{12}) \div (3.2 \times 10^{-6})$
   b) $(7.4 \times 10^9)^2 + (4.1 \times 10^{11})$

SURDS

Rational numbers include $\frac{1}{3}$, $0.\dot{7}$, $\sqrt{16}$, $\sqrt[3]{8}$ etc.

Irrational numbers include π, π^2, $\sqrt{2}$, $\sqrt{7}$ etc.

Irrational numbers involving square roots are also called surds.

Manipulating surds

When working with surds there are several rules to learn:

SURDS

1. $\sqrt{a} \times \sqrt{b} = \sqrt{ab}$
   e.g. $\sqrt{3} \times \sqrt{5} = \sqrt{15}$

2. $(\sqrt{b})^2 = \sqrt{b} \times \sqrt{b} = b$
   eg $(\sqrt{5})^2 = \sqrt{5} \times \sqrt{5} = 5$

3. $\frac{\sqrt{a}}{\sqrt{b}} = \sqrt{\frac{a}{b}}$
   $\frac{\sqrt{10}}{\sqrt{2}} = \sqrt{\frac{10}{2}} = \sqrt{5}$

4. $(a + \sqrt{b})^2$
   $= (a + \sqrt{b})(a + \sqrt{b})$
   $= a^2 + 2a\sqrt{b} + (\sqrt{b})^2$
   $= a^2 + 2a\sqrt{b} + b$

5. $(a + \sqrt{b})(a - \sqrt{b})$
   $= a^2 - a\sqrt{b} + a\sqrt{b} - (\sqrt{b})^2$
   $= a^2 - b$

Examples

1. Simplify $\sqrt{75}$
   $\sqrt{75} = \sqrt{25} \times \sqrt{3}$
   $\sqrt{75} = 5\sqrt{3}$

   Look for perfect square, i.e. 25.

2. Simplify $(\sqrt{3} + 2)^2$

   $(\sqrt{3} + 2)(\sqrt{3} + 2)$

   $= \sqrt{9} + 2\sqrt{3} + 2\sqrt{3} + 4$

   $= 7 + 4\sqrt{3}$

3. Work out $\frac{(2 - \sqrt{2})(4 + 3\sqrt{2})}{2}$.

   Leave your answer in surd form.

   $\frac{8 + 6\sqrt{2} - 4\sqrt{2} - 3(\sqrt{2})^2}{2}$

   $\frac{8 + 2\sqrt{2} - 6}{2}$

   $\frac{2 + 2\sqrt{2}}{2}$

   $= \frac{2(1 + \sqrt{2})}{2}$

   $= 1 + \sqrt{2}$

A rational number is one that can be expressed in the form $\frac{a}{b}$, where a and b are integers.

15 MINS

DAY 1 2 3 4 5 6 7

Rationalising the denominator

Sometimes a surd can appear on the bottom of the fraction. It is usual to rewrite the surd so that it appears as the numerator. This is called **rationalising** the denominator.

Example

Rationalise $\frac{4}{\sqrt{7}}$

$= \frac{4}{\sqrt{7}} \times \frac{\sqrt{7}}{\sqrt{7}}$

Multiply the top and bottom of the fraction by the surd function. In this case $\sqrt{7}$.

$= \frac{4\sqrt{7}}{(\sqrt{7})^2}$

$= \frac{4\sqrt{7}}{7}$

Progress check

1. Express the following in the form $a\sqrt{b}$.
   a) $\sqrt{24}$
   b) $\sqrt{200}$
   c) $\sqrt{48} + \sqrt{12}$

2. Molly works out $(2 - \sqrt{3})^2$. She says the answer is 1. Decide whether Molly is correct, giving a reason for your answer.

3. Work out $\frac{(5 + \sqrt{5})\,(2 - 2\sqrt{5})}{\sqrt{45}}$.

   Give your answer in its simplest form.

4. Decide whether this is correct.

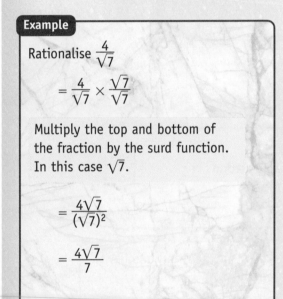

   $\frac{1}{\sqrt{2}} = \frac{\sqrt{2}}{2}$

PROPORTIONALITY

Direct proportion

As one variable increases, the other increases and as one variable decreases the other decreases.

Example

If a is proportional to the square of b and $a = 5$ when $b = 4$, find the value of k (the constant of proportionality) and the value of a when $b = 8$.

STEP 1

Change the sentence into proportionality using the symbol $\propto$.

$a \propto b^2$

STEP 2

Replace $\propto$ with "$= k$" to make an equation.

$a = kb^2$

STEP 3

Substitute the value given in the question in order to find k. Rearrange the equation $5 = k \times 4^2$

$\frac{5}{16} = k$

STEP 4

Replace k with the value just found.

$a = \frac{5}{16}b^2$

If $b = 8$

$a = \frac{5}{16} \times 8^2$

$a = \frac{5}{16} \times 64$

$= 20$

The notation $\propto$ means 'is directly proportional to'. This is often abbreviated to 'is proportional to' or 'varies as'.

20 MINS

DAY 2

Inverse proportion

- As one variable increases the other decreases, and as one variable decreases the other increases.
- If y is inversely proportional to x then we write this as $y \propto \frac{1}{x}$ or else $y = \frac{k}{x}$.

Example

p is inversely proportional to the cube of w. If $w = 2$ when $p = 5$, what is the value of w when $p = 10$?

$p \propto \frac{1}{w^3}$	Write the information with the proportionality sign.
$p = \frac{k}{w^3}$	Replace with the constant of proportionality.
$5 = \frac{k}{2^3}$	
$5 \times 8 = k$	
$k = 40$	Find the value of k.
$p = \frac{40}{w^3}$	Rewrite the equation.
$10 = \frac{40}{w^3}$	find the value of w if $p = 10$.
$w^3 = 4$	
$w = \sqrt[3]{4}$	
$w = 1.587$	

Progress check

1 Match these statements with the correct equation.

a) y is proportional to x	$y = \frac{k}{x}$
b) y is inversely proportional to cube root of x	$y = kx^3$
c) y is inversely proportional to x	$y = kx$
d) y is proportional to x^3	$y = \frac{k}{\sqrt[3]{x}}$

2 a is proportional to $\sqrt{x}$. When $x = 4$, $a = 8$. What is the equation of proportionality and what is the value of x when $a = 64$?

$a \propto \sqrt{x}$ $a = k\sqrt{x}$

$a\sqrt{x} = k$

$8 \times \sqrt{4} = k = 8 \times 2$

$k = 16$

$a = 16\sqrt{x}$

if $a = 64$ $\frac{64}{16} = 4$

$64 = 16\sqrt{16}$

$\frac{a}{16} = \sqrt{x}$

$\sqrt{x} = 16$ $x = 256$

UPPER AND LOWER BOUNDS OF MEASUREMENT

When measurements are quoted to a given unit, say nearest metre, there is a highest and lowest value they could be.

Highest value – called the Upper bound

Lowest value – called the Lower bound

Example

6.2 cm rounded to the nearest millimetre would really lie between

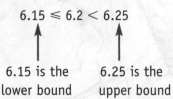

$$6.15 \leqslant 6.2 < 6.25$$

6.15 is the lower bound ↑ 6.25 is the upper bound ↑

The real value can be as much as half the rounded unit below or above the rounded off value.

Finding maximum and minimum possible values of a calculation

Some exam questions may ask you to calculate the upper or lower bounds of calculations.

TIPS

	Upper bound	**Lower bound**
Addition	Upper bound + Upper bound	Lower bound + Lower bound
Multiplication	Upper bound × Upper bound	Lower bound × Lower bound
Subtraction	Upper bound of larger quantity − Lower bound of smaller quantity	Lower bound of larger quantity − Upper bound of smaller quantity
Division	$\frac{\text{Upper bound of quantity 1}}{\text{Lower bound of quantity 2}}$	$\frac{\text{Lower bound of quantity 1}}{\text{Upper bound of quantity 2}}$

Measurements are never exact. They can only be expressed to a certain degree of accuracy.

15 MINS

Examples

1. A square measures 62 mm to the nearest mm. Work out the upper and lower area of the square (in mm^2).

62 → 61.5 (lower bound), 62.5 (upper bound)

Work out the upper and lower bounds for 62 mm.

Upper bound $= 62.5 \times 62.5$

$= 3906.25\,mm^2$

Lower bound $= 61.5 \times 61.5$

$= 3782.25\,mm^2$

2. Given that $a = 4.2$ (to 1dp) and $b = 6.23$ (to 3sf), find the upper and lower bounds for the following calculations:

a) $b - a$ b) $\frac{b}{a}$

a) 4.2 → 4.15, 4.25   6.23 → 6.225, 6.235

a) Upper bound $= 6.235 - 4.15 = 2.085$

Lower bound $= 6.225 - 4.25 = 1.975$

b) Upper bound $= \frac{6.235}{4.15} = 1.5024$

Lower bound $= \frac{6.225}{4.25} = 1.4647$

Progress check

1 $c = \frac{(2.6)^3 \times 12.52}{3.2}$

2.6 and 3.2 are correct to 1 decimal place. 12.52 is correct to 2 decimal places. Which of the following calculations give the lower bound for c and the upper bound for c?

A $\frac{(2.65)^3 \times 12.525}{3.15}$

B $\frac{(2.55)^3 \times 12.515}{3.15}$

C $\frac{(2.65)^3 \times 12.525}{3.25}$

D $\frac{(2.65)^3 \times 12.515}{3.15}$

E $\frac{(2.55)^3 \times 12.515}{3.25}$

2 Work out the upper and lower bound of $\frac{6}{p}$, where p is 27 (2sf).

DAY 2

FORMULAE

15 MINS

Substituting into formulae

Replacing a letter with a number is called **substitution.**

- **Write out the expression first and then replace the letters with the values given.**
- **Work out the value – but take care with the order of operations, i.e. BIDMAS.**

Examples

1. $a = 3b - 4c$

   Find a if $b = 4$ and $c = -2$.

   $a = (3 \times 4) - (4 \times -2)$

   $= 12 - (-8)$

   $= 20$

   Taking away a negative is the same as adding.

2. $E = \frac{1}{2}mv^2$

   Find E if $m = 6$ and

   $v = 10$.

   $= \frac{1}{2} \times 6 \times 10^2$

   $\frac{1}{2}$ $=$ $\times 6 \times$

   E $= 3$

Rearranging formulae

- The subject of a formula is the letter that appears on its own on one side of the formula.

Example

Make a the subject of the formula $b = (a - 3)^2$.

$b = (a - 3)^2$ Deal with the power first, square root both sides.

$\pm\sqrt{b} = a - 3$ Remove any term added or subtracted. Add 3 to both sides.

$\pm\sqrt{b} + 3 = a$

$a = \pm\sqrt{b} + 3$ or $a = 3 \pm \sqrt{b}$

DAY 2

$b = a + 6$ is called a formula. The value of b depends on the value of a.

Example

Sometimes the subject appears in more than one term.

Make x the subject of the formula

$$a = \frac{x + c}{x - d}.$$

$a = \dfrac{x + c}{x - d}$	Multiply both sides by $(x - d)$.
$a(x - d) = x + c$	Multiply out the brackets.
$ax - ad = x + c$	Collect like terms involving x on one side of the equation.
$ax - x = c + ad$	
$x(a - 1) = c + ad$	Factorise.
$x = \dfrac{c + ad}{(a - 1)}$	

These types of questions are usually worth at least 3 marks.

Progress check

1. If $a = \frac{3}{5}$ and $b = -2$, find the value of these expressions, giving your answer to 3sf where appropriate
   a) $ab - 5$
   b) $a^2 + b^2$
   c) $3a - 6ab$

2. Make u the subject of the formula $v^2 = u^2 + 2as$.

3. Make p the subject of the formula $q = \dfrac{p - t}{p + v}$.

BRACKETS AND FACTORISATION

Single brackets

Each term outside the brackets multiplies each separate term inside the bracket.

Examples

Expand and simplify:

a) $5(x + 6) = 5x + 30$

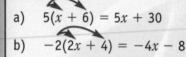

b) $-2(2x + 4) = -4x - 8$

c) $3(2x - 5) - 2(x - 3)$

$= 6x - 15 - 2x + 6$ Multiply out the brackets.

$= 4x - 9$ Collect like terms.

Two brackets

$(x + 4)(x + 2) = x^2 + 2x + 4x + 8$

$= x^2 + 6x + 8$

- Every term in the second bracket must be multiplied by every term in the first bracket.
- Often, but not always, the two middle terms are like terms and can be collected together.

Examples

$(x + 4)(2x - 5) = 2x^2 - 5x + 8x - 20$

$= 2x^2 + 3x - 20$

$(2x + 1)^2 = (2x + 1)(2x + 1)$

$= 4x^2 + 2x + 2x + 1$

$= 4x^2 + 4x + 1$ Remember that x^2 means x times by itself.

Multiplying out brackets helps to simplify algebraic expressions.

10 MINS

Factorisation

Factorisation simply means putting into brackets.

One bracket

$4x + 6 = 2(2x + 3)$

To factorise $4x + 6$

- Recognise that 2 is the highest common factor of 4 and 6.
- Take out the common factor.
- The expression is completed inside the bracket so that when multiplied out it is equivalent to $4x + 6$.

Two brackets

Two brackets are obtained when a quadratic expression of the type $ax^2 + bx + c$ is factorised.

Examples

$$x^2 + 4x + 3 = (x + 1)(x + 3)$$
$$x^2 - 7x + 12 = (x - 3)(x - 4)$$
$$x^2 + 3x - 10 = (x + 5)(x - 2)$$

Progress check

1. Expand and simplify:
   a) $(x + 3)(x - 2)$
   b) $2(3x - 4)$
   c) $4x(x - 3)$
   d) $(x - 3)^2$

2. Factorise:
   a) $4x^2 + 8x$
   b) $12xy - 6x^2$
   c) $3a^2b + 6ab^2$

3. Factorise:
   a) $x^2 + 4x + 4$
   b) $x^2 - 5x + 6$
   c) $x^2 - 4x - 5$

EQUATIONS 1

Linear equations

Type 1 of the form $ax + b = c$

Remember to do the same thing to both sides of the equation so that they balance.

Example

Solve:

$$5x - 2 = 13$$

$5x = 13 + 2$ — Add 2 to both sides.

$5x = 15$

$x = 15 \div 5$ — Divide both sides by 5.

$x = 3$

Type 2 of the form $ax + b = cx + d$

Example

Solve:

$$7x - 4 = 3x + 8$$

$7x = 3x + 12$ — Add 4 to each side.

$4x = 12$ — Subtract $3x$ from both sides.

$x = 12 \div 4$

$x = 3$

Must check:

$7 \times 3 - 4 = 3 \times 3 + 8$

$21 - 4 = 9 + 8$

Yes, it's right. ✔

Type 3 with brackets!

Examples

(i) Solve: $5(x - 1) = 3(x + 2)$

$5x - 5 = 3x + 6$

$5x = 3x + 11$

$2x = 11$

$x = 11 \div 2$

$x = 5.5$

Just multiply out the brackets and solve as normal.

(ii) Solve: $\frac{3(2x - 1)}{5} = 6$

Multiply both sides by 5.

$3(2x - 1) = 6 \times 5$

$6x - 3 = 30$

$6x = 33$

$x = 33 \div 6$

$x = 5.5$

Type 4 problems

Always write down the information that you know.

The perimeter of this rectangle is 30 cm. Work out the value of y and find the length of the rectangle.

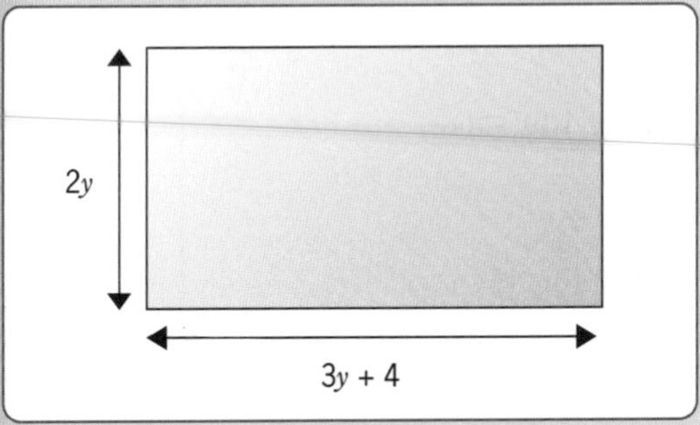

Write down what you know.

$3y + 4 + 2y + 3y + 4 + 2y = 30$

Simplify the expression and solve as normal.

$10y + 8 = 30$

$10y = 30 - 8$

$10y = 22$

$y = 2.2$

Length of rectangle $= 3 \times 2.2 + 4$

$= 10.6$ cm

EQUATIONS 2

Solving equations involving indices

Equations sometimes involve indices. You need to remember the laws of indices to be able to solve them.

Example

Solve:

1. $y^k = \sqrt[3]{y} \div \frac{1}{y^5}$

$y^k = y^{\frac{1}{3}} \div y^{-5}$ Rewrite as indices.

$y^k = y^{\frac{1}{3} - (-5)}$ When dividing subtract the indices.

$y^k = y^{5\frac{1}{3}}$

$k = 5\frac{1}{3}$ Compare the indices since the bases are the same.

Example

Solve:

$2^{k+2} = 32$

$2^{k+2} = 2^5$ Rewrite so that both bases are the same.

$k+2 = 5$

$k = 3$

Example

The area of this rectangle is 81 cm^2

Work out the value of k.

$3^{k+2} \times \sqrt{3} = 81$	Write out the equation.
$3^{k+2} \times 3^{\frac{1}{2}} = 81$	
$3^{(k+2+\frac{1}{2})} = 3^4$	Write out so that the bases are 3.
$3^{(k+\frac{5}{2})} = 3^4$	Add the indices of the LHS.
$k + \frac{5}{2} = 4$ $k = \frac{3}{2}$	Compare the indices.

Progress check

Solve the following equations:

1. $2x - 6 = 10$
2. $5 - 3x = 20$
3. $4(2 - 2x) = 12$
4. $6x + 3 = 2x - 10$
5. $7x - 4 = 3x - 6$
6. $5(x + 1) = 3(2x - 4)$
7. The perimeter of this triangle is 60 cm. Work out the value of x and find the shortest length.

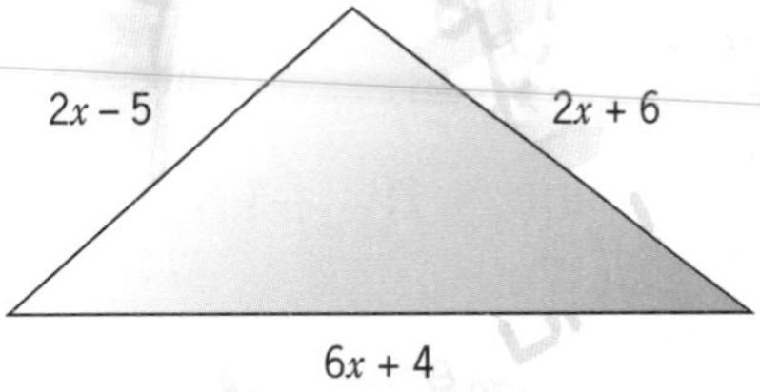

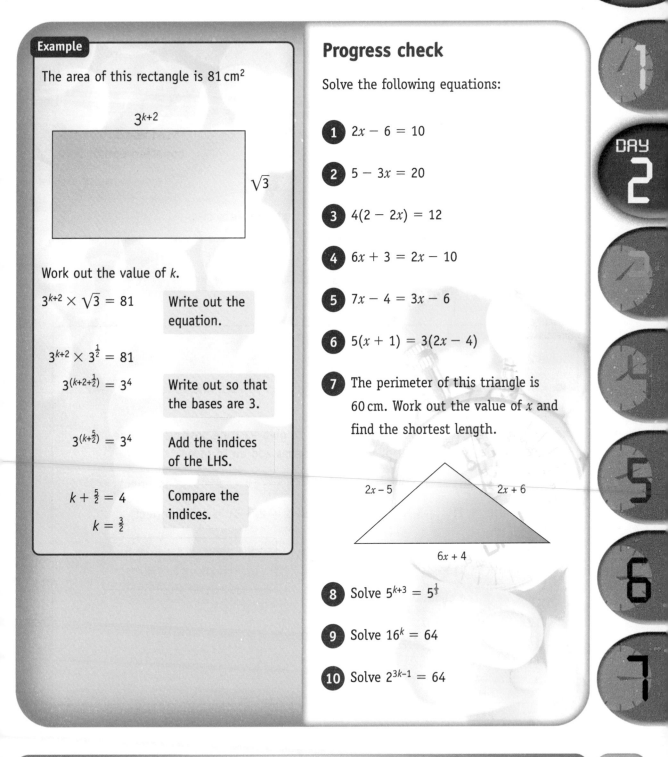

8. Solve $5^{k+3} = 5^{\frac{1}{3}}$
9. Solve $16^k = 64$
10. Solve $2^{3k-1} = 64$

SOLVING QUADRATIC AND CUBIC EQUATIONS 1

Solving quadratic equations

Solve: $x^2 - x - 6 = 0$

Must check that the equation equals zero.

It's a quadratic of the form $ax^2 + bx + c = 0$

Need to factorise into two brackets (   )(   ) = 0

Method 1: Factorisation

Example

Solve $2x^2 - x - 3 = 0$.

- Write out the two brackets and put an x and $2x$ in each one, since $x \times 2x = 2x^2$.
  $(2x \quad)(x \quad)$
- We now need two numbers that multiply to give –3 (one positive and one negative) and when multiplied by the x terms add up to give $-1x$.

  $(2x + 1)(x - 3)$ $1 \times -3 = -3$ ✔ x-terms $= -6x + x = 5x$ ✗ incorrect

  $(2x - 3)(x + 1)$ $-3 \times 1 = -3$ ✔ x-terms $= 2x - 3x = -x$ ✔ correct

  A quick check gives
  $(2x - 3)(x + 1) = 2x^2 - x - 3$

  The $2x$ and 1 must be in different brackets.

- Now solve the equation:

  $(2x - 3)(x + 1) = 0$
  $\therefore$ $(2x - 3) = 0$ so $x = \frac{3}{2}$
  or $(x + 1) = 0$ so $x = -1$
  $\therefore$ $x = \frac{3}{2}$ or -1

Quadratic equations can be solved using a variety of methods.

10 MINS

Method 2 The quadratic formula

Where a quadratic does not factorise then you can use the formula:

$$x = \frac{-b \pm \sqrt{b^2 - 4ac}}{2a}$$

for any quadratic equation written in the form $ax^2 + bx + c = 0$
This formula will be given on the exam paper.

Example

Solve the equation $2x^2 - 7x = 5$. Give your answers to 2 decimal places.

a) Put the equation into the form $ax^2 + bx + c = 0$: $2x^2 - 7x - 5 = 0$

b) Identify the values of a, b and c: $a = 2$, $b = -7$, $c = -5$

c) Substitute these values into the quadratic formula:

$$x = \frac{-b \pm \sqrt{b^2 - 4ac}}{2a}$$

$$x = \frac{7 \pm \sqrt{(-7)^2 - (4 \times 2 \times -5)}}{2 \times 2}$$

$$x = \frac{7 \pm \sqrt{49 - (-40)}}{4}$$

$$x = \frac{7 \pm \sqrt{89}}{4}$$

$$x = \frac{7 + \sqrt{89}}{4}$$

One solution is when we use $+\sqrt{89}$

$x = 4.11$ (2dp)

$$x = \frac{7 - \sqrt{89}}{4}$$

One solution is when we use $-\sqrt{89}$

$x = -0.61$ (2dp)

Check

$2 \times (4.11)^2 - 7(4.11) - 5 = 0$ ✔

$2 \times (-0.61)^2 - 7(-0.61) - 5 = 0$ ✔

DAY 2

SOLVING QUADRATIC AND CUBIC EQUATIONS 2

Completing the square

This is when quadratic equations are expressed in the form

$$(x + a)^2 + b = 0$$

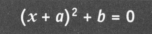

- **Rearrange the equation into the form $ax^2 + bx + c = 0$**
- **If a is not 1, divide the whole equation by a.**
- **Write the equation in the form $(x + \frac{b}{2})^2$. Notice that the number in the bracket is always half the value of b.**
- **Multiply out the brackets, compare to the original and adjust by adding or subtracting an extra amount.**

Example

Express $x^2 - 6x + 2 = 0$ as a completed square and hence solve it.

$x^2 - 6x + 2 = 0$	Rearrange in the form $ax^2 + bx + c = 0$.
$(x - 3)^2$	Half of –6 is –3.
$(x - 3)^2 = x^2 - 6x + 9$	Multiply out the brackets and now compare to the original. $x^2 - 6x + 9$
$(x - 3)^2 - 7 = 0$	To make $(x - 3)^2$ like the original we subtract 7.

Now solve:

$(x - 3)^2 = 7$

$(x - 3) = \pm \sqrt{7}$

$\therefore \quad x = \sqrt{7} + 3$ or $x = -\sqrt{7} + 3$

$x = 5.65$ or $x = 0.35$

Solving cubic equations by trial and improvement

Trial and improvement gives an approximate solution to cubic equations.

Example

The equation $x^3 + 2x = 58$ has a solution between 3 and 4. Find the solution to 1 decimal place.

Drawing a table can help you – but also the examiner – since it makes it easier to follow what you have done.

x	$x^3 + 2x$	Comment
3.5	$3.5^3 + 2 \times 3.5 = 49.875$	too small
3.8	$3.8^3 + 2 \times 3.8 = 62.472$	too big
3.7	$3.7^3 + 2 \times 3.7 = 58.053$	too big
3.65	$3.65^3 + 2 \times 3.65 = 55.927$	too small
		$x = 3.7$ (1dp)

Progress check

1. The equation $a^3 = 40 - a$ has a solution between 3 and 4. Find the solution to 1dp, by using a method of trial and improvement.

2. Solve these equations by (i) factorisation (ii) using the quadratic formula.
   a) $x^2 + 2x - 15 = 0$
   b) $2x^2 + 5x + 2 = 0$

3. The expression $x^2 + 8x + 2$ can be written in the form $(x + a)^2 + b$ for all values of x.
   a) Find a and b.
   b) The expression $x^2 + 8x + 2$ has a minimum value. Find this minimum value.

SIMULTANEOUS LINEAR EQUATIONS

By algebra (elimination method)

Solve simultaneously:	$3x + 2y = 8$ $2x - 3y = 14$
Label the equations ① and ②.	$3x + 2y = 8$ ① $2x - 3y = 14$ ②
Since no coefficients match, multiply equation ① by 2 and equation ② by 3.	$6x + 4y = 16$ $6x - 9y = 42$
Rename them equations ③ and ④.	$6x + 4y = 16$ ③ $6x - 9y = 42$ ④
The coefficient of x in equations ③ and ④ is the same. Subtract equation ④ from equation ③ and solve remaining equation.	$0x + 13y = -26$ $y = -26/13$ $y = -2$
Substitute the value of $y = -2$ back into equation ①. Solve this equation to find x.	$3x + (-4) = 8$ $3x = 8 + 4$ $3x = 12$ $x = 4$
Check in equation ②.	$(2 \times 4) - (3 \times -2) = 14$ ✔
Solution is: $x = 4$, $y = -2$	

Two equations with two unknowns are called simultaneous equations. They can be solved graphically or algebracially.

15 MINS

Graphically

Solve the simultaneous equations

$2x + 3y = 6$
$x + y = 1$

The point at which any two graphs intersect represents the simultaneous solutions of their equations.

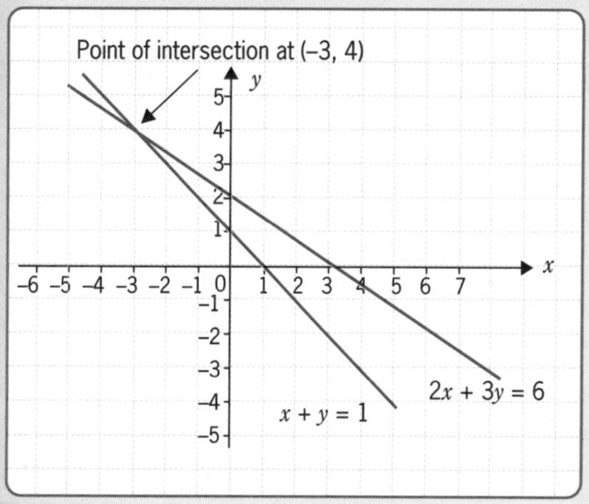

Draw the graph of:

$2x + 3y = 6$
$x = 0\ 3y = 6 \therefore y = 2\ (0, 2)$
$y = 0\ 2x = 6 \therefore x = 3\ (3, 0)$

Draw the graph of:

$x + y = 1$
$x = 0, y = 1 \therefore (0, 1)$
$y = 0, x = 1 \therefore (1, 0)$

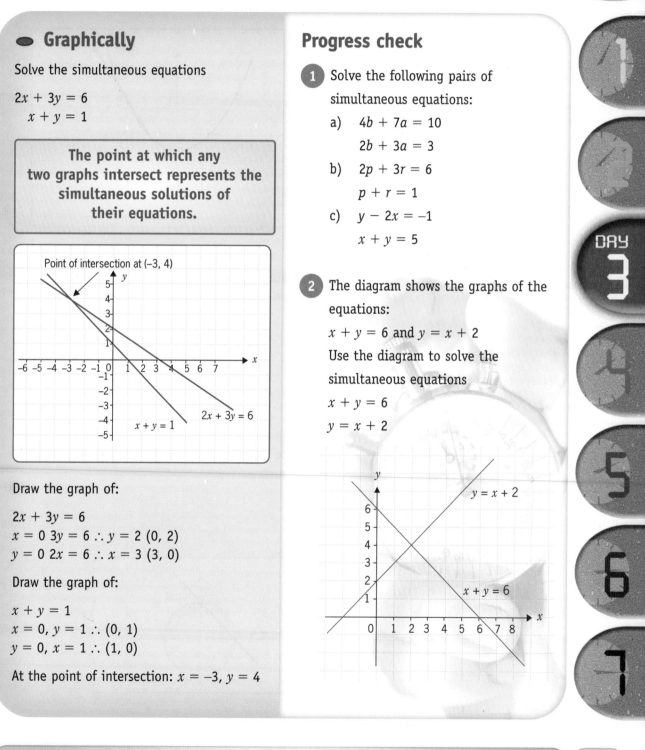

At the point of intersection: $x = -3$, $y = 4$

Progress check

1. Solve the following pairs of simultaneous equations:
   a) $4b + 7a = 10$
      $2b + 3a = 3$
   b) $2p + 3r = 6$
      $p + r = 1$
   c) $y - 2x = -1$
      $x + y = 5$

2. The diagram shows the graphs of the equations:
   $x + y = 6$ and $y = x + 2$
   Use the diagram to solve the simultaneous equations
   $x + y = 6$
   $y = x + 2$

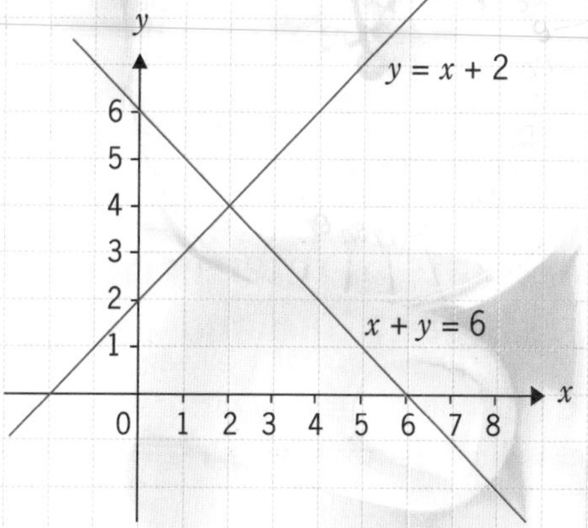

SOLVING LINEAR AND QUADRATIC EQUATIONS SIMULTANEOUSLY

Point of intersection of a straight line and a curved graph

The diagram shows the points of intersection A and B of the line $y = 4x - 2$ and the curve $y = x^2 + 1$. We know that the points A and B lie on both the curve and the line

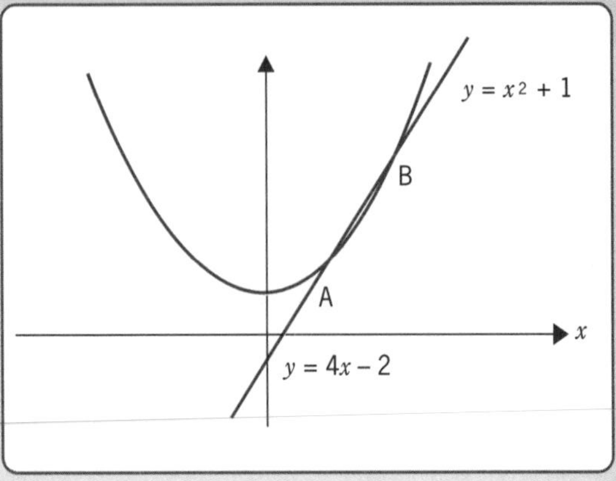

$y = 4x - 2$ ①

$y = x^2 + 1$ ②

Eliminate y by substituting equation ② into equation ①.

$x^2 + 1 = 4x - 2$

rearrange $x^2 - 4x + 3 = 0$

factorise $(x - 1)(x - 3) = 0$

so $x = 1$ and $x = 3$

when $x = 1$, $y = 1^2 + 1 = 2$

Now substitute the values of x back into equation ② to find the corresponding values of y.

when $x = 3$, $y = 3^2 + 1 = 10$

Coordinates of A are (1, 2).
Coordinates of B are (3, 10).

Points of intersection of a straight line and a circle

The equation of a circle with centre (0, 0) and radius r is $\mathbf{x^2 + y^2 = r^2}$.

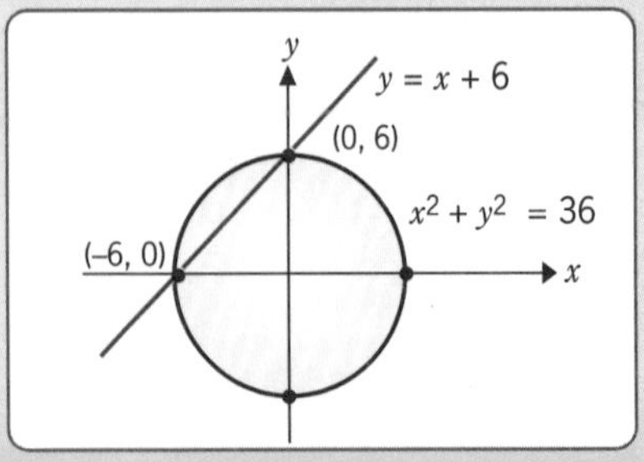

You need to be able to work out the coordinates of the points of intersection of a straight line and a quadratic curve as well as a straight line and a circle.

15 MINS

Find the coordinates of the points where the line $y - x = 6$ cuts $x^2 + y^2 = 36$.

The coordinates must satisfy both equations so we solve them simultaneously.

$y - x = 6$ ① $x^2 + y^2 = 36$ ② $y = x + 6$ ①	Rewrite equation ① in the form '$y =$'.

Eliminate y by substituting equation ① into equation ②

$x^2 + (x + 6)^2 = 36$	Expand brackets.

$$(x + 6)^2 = (x + 6)(x + 6) = x^2 + 12x + 36$$

$x^2 + x^2 + 12x + 36 = 36$

$2x^2 + 12x = 0$	Rearrange.
$2x(x + 6) = 0$	Factorise.
$x = 0, x = -6$	Solve.

Substitute the values of x into equation ① to find the values of y.

$x = 0, y = 0 + 6 = 6$

$x = -6, y = -6 + 6 = 0$

The line $y - x = 6$ cuts the circle $x^2 + y^2 = 36$ at (0, 6) and (–6, 0).

Progress check

1 Solve these simultaneous equations

a) $y = x + 4$
   $y = x^2 + 2$

b) $y = x + 1$
   $x^2 + y^2 = 25$

2 For the two questions above give a geometrical interpretation of the result.

ALBEBRAIC FRACTIONS

15 MINS

DAY 3

Simplifying algebraic fractions

Example

Simplify $\frac{8x + 16}{x^2 - 4}$.

- Firstly you need to factorise the numerator and denominator.

$$\frac{8(x + 2)}{(x + 2)(x - 2)}$$

Remember the difference of two squares: $(b^2 - a^2) = (b - a)(b + a)$

- Now cancel any common factors, i.e. the $(x + 2)$.

$$\frac{8\cancel{(x + 2)}}{\cancel{(x + 2)}(x - 2)} = \frac{8}{(x - 2)}$$

Multiplication

- Multiply the numerators together and the denominators together.
- Cancel if possible.

Example

$$\frac{3a^2}{4b} \times \frac{16b^2}{9ab} = \frac{48a^2b^2}{36ab^2} = \frac{4}{3}a$$

Division

- Remember to turn the second fraction upside down (i.e. take the reciprocal) then multiply and cancel if possible.

When working with algebraic fractions, the same rules are used as with ordinary fractions.

Example

$$\frac{12(x-3)}{(x+2)} \div \frac{4(x-1)(x-3)}{(x+1)}$$

Cancel out common factors.

$$= \frac{{}^{3}\cancel{12}\cancel{(x-3)}}{(x+2)} \times \frac{(x+1)}{\cancel{4}(x-1)\cancel{(x-3)}}$$

$$= \frac{3(x+1)}{(x-1)(x+2)}$$

Addition and subtraction

- As with ordinary fractions, you cannot add or subtract unless the fractions have the same denominator.

Example

$$\frac{(x+3)}{(x+2)} + \frac{3}{(x-1)}$$

- The common denominator is: $(x+2)(x-1)$.
- The numerator now needs to be adjusted:

$$\frac{(x+3)(x-1)+3(x+2)}{(x+2)(x-1)}$$

- Multiply out the numerator:

$$\frac{x^2+2x-3+3x+6}{(x+2)(x-1)}$$

- Simplify, then factorise the numerator if possible and then simplify by cancelling:

$$= \frac{x^2+5x+3}{(x+2)(x-1)}$$

Progress check

1. Decide whether these expressions are fully simplified:

a) $\frac{6s^2w}{3w^2}$

b) $\frac{5a^2b}{2c}$

c) $\frac{9abc}{d}$

2. Simplify the following algebraic fractions:

a) $\frac{3}{(x+1)} + \frac{2}{(x-1)}$

b) $\frac{3a^2b}{2q} \times \frac{4q^2}{9abc}$

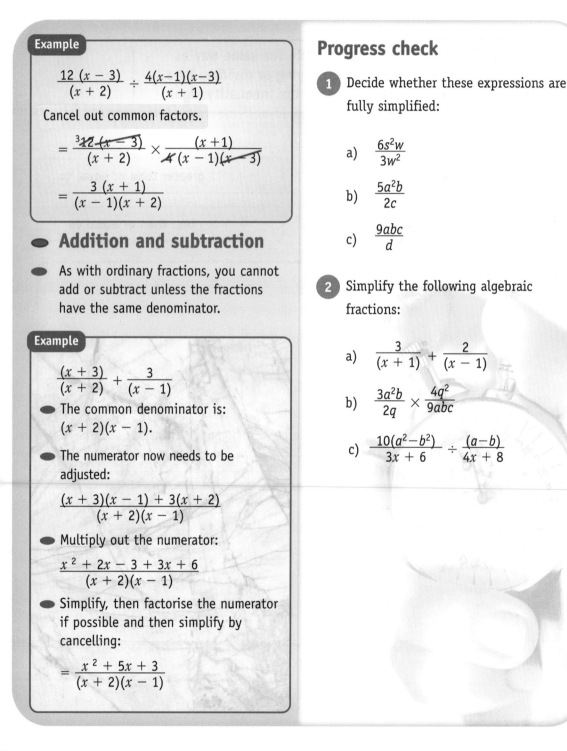

c) $\frac{10(a^2-b^2)}{3x+6} \div \frac{(a-b)}{4x+8}$

INEQUALITIES

Inequalities can be solved in exactly the same way as equations except that when multiplying or dividing by a negative number, you must reverse the inequality sign.

The inequality symbols

$>$ means **greater than**

$\geqslant$ means **greater than or equal to**

$<$ means **less than**

$\leqslant$ means **less than or equal to**

Example

Solve:

$2x - 2 < 10$

$2x < 10 + 2$

$2x < 12$

$x < 6$

Solve:

$3 - 2x \geqslant 9$

$-2x \geqslant 9 - 3$

$-2x \geqslant 6$

$x \leqslant \frac{6}{-2}$

$x \leqslant -3$ Divide by –2 and change inequality sign round.

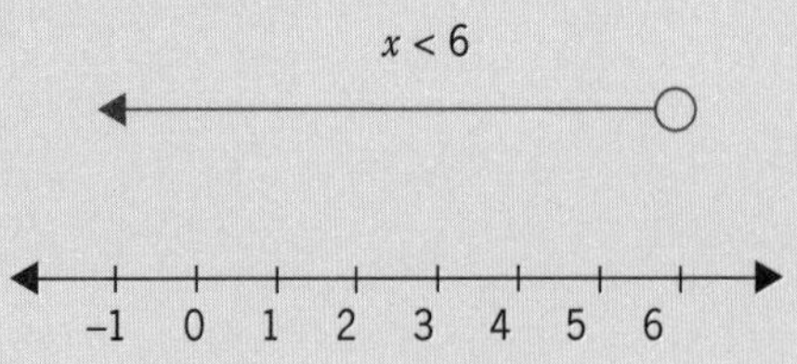

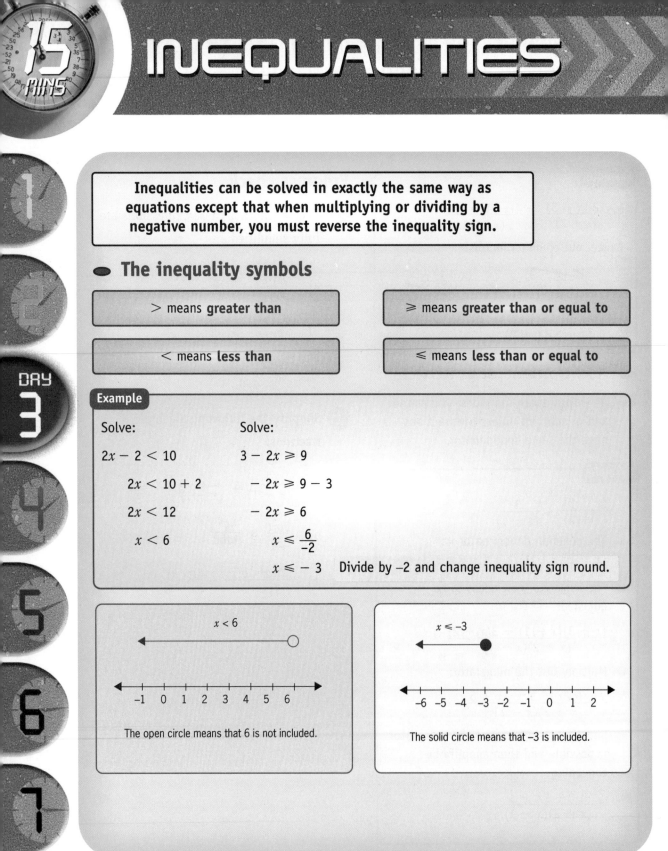

The open circle means that 6 is not included.

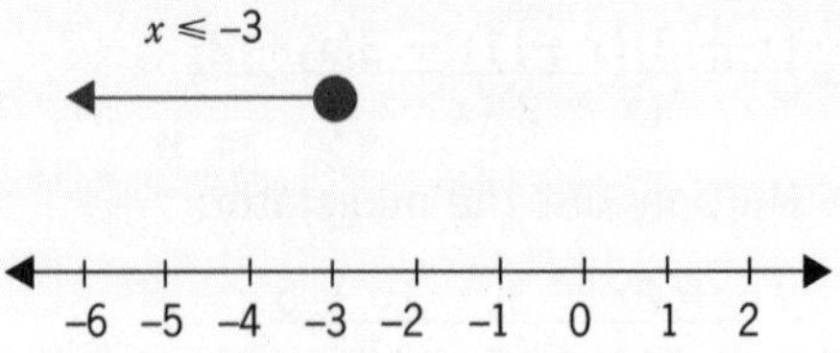

The solid circle means that –3 is included.

In inequalities the LHS does not equal the RHS.

15 MINS

Example

Solve:

$-2 < 4x - 3 \leqslant 9$
$1 < 4x \leqslant 12$
$\frac{1}{4} < x \leqslant 3$

The integer values that satisfy this inequality are 1, 2, 3.

Graphs of inequalities

The graph of an equation such as $x = 2$ is a line, whereas the graph of the inequality $x < 2$ is a region that has $x = 2$ as its boundary.

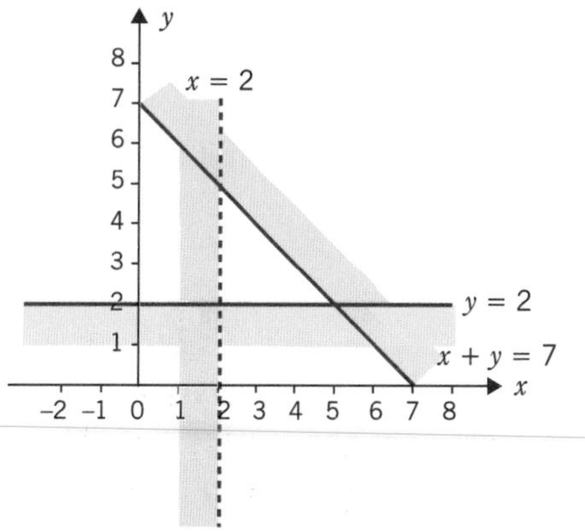

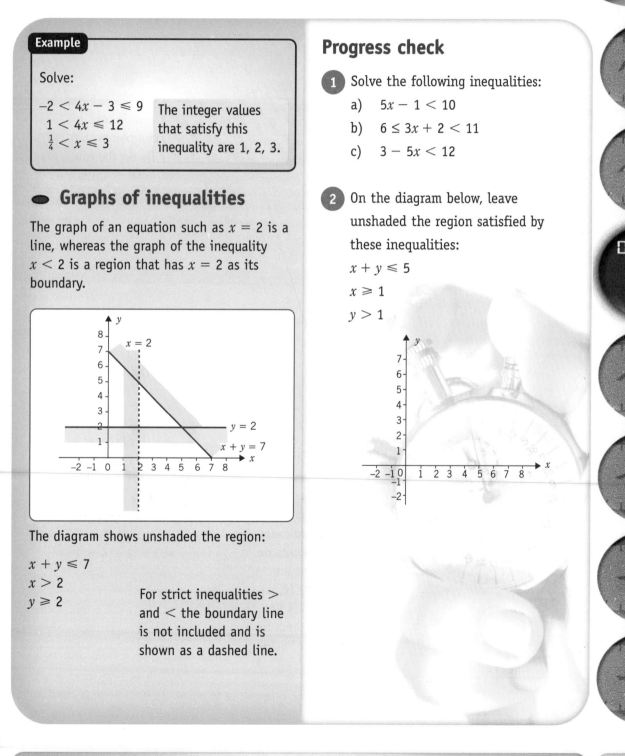

The diagram shows unshaded the region:

$x + y \leqslant 7$
$x > 2$
$y \geqslant 2$

For strict inequalities > and < the boundary line is not included and is shown as a dashed line.

Progress check

1. Solve the following inequalities:
   a) $5x - 1 < 10$
   b) $6 \leq 3x + 2 < 11$
   c) $3 - 5x < 12$

2. On the diagram below, leave unshaded the region satisfied by these inequalities:
   $x + y \leqslant 5$
   $x \geqslant 1$
   $y > 1$

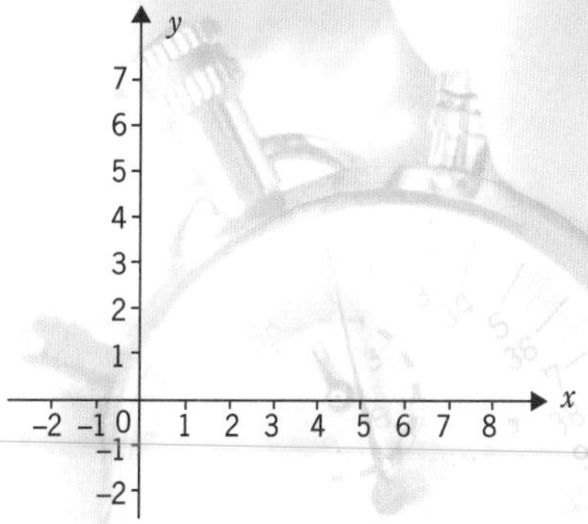

DAY 3

STRAIGHT-LINE GRAPHS

1 To work out the coordinates of the points that lie on the line $y = 3x - 4$, draw a table of values.

x	−1	0	2	4
y	−7	−4	2	8

2 Substitute the x values into the equation $y = 3x - 4$, to find the values of y i.e. $x = 2$, $y = 3 \times 2 - 4 = 2$

3 The coordinates of the points on the line are: (−1, −7) (0, 4) (2, 2) (4,8) Just read them from the table of values.

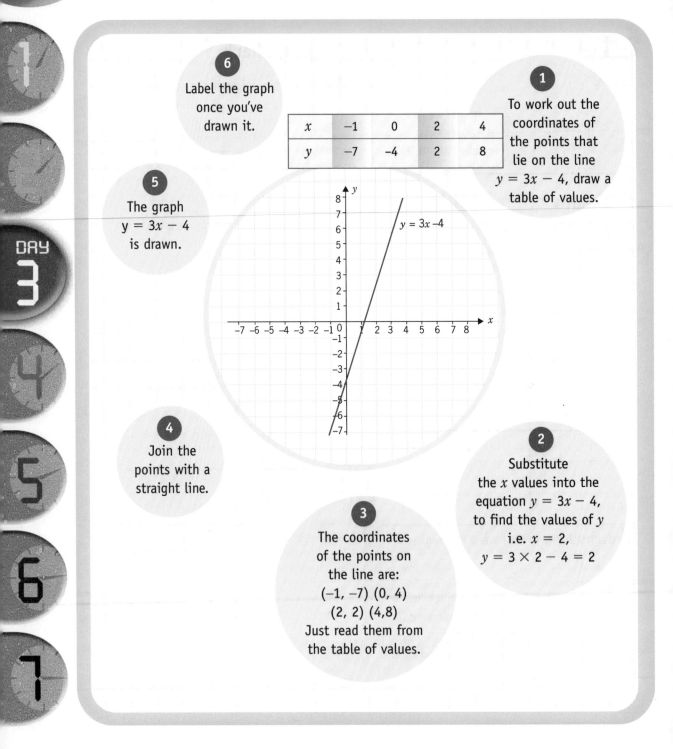

4 Join the points with a straight line.

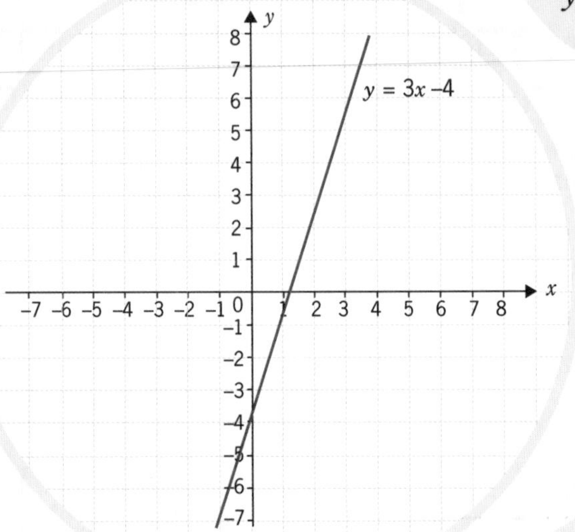

5 The graph y = $3x - 4$ is drawn.

6 Label the graph once you've drawn it.

The general equation of a straight lined graph is $y = mx + c$

m is the gradient. c is the intercept on the y-axis.

Gradient of a straight line

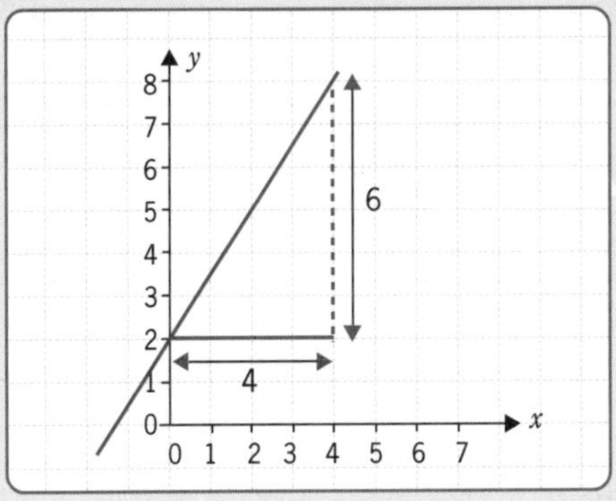

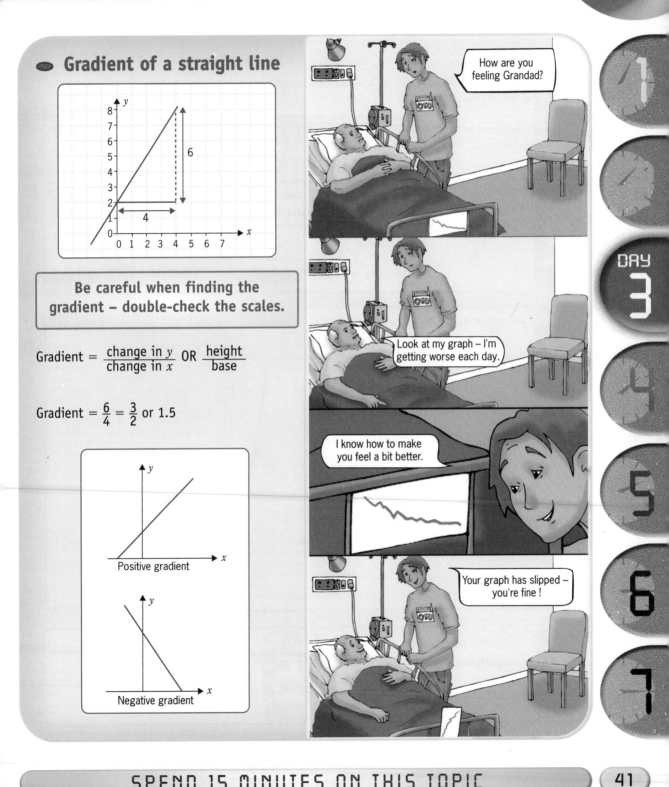

Be careful when finding the gradient – double-check the scales.

Gradient $= \dfrac{\text{change in } y}{\text{change in } x}$ OR $\dfrac{\text{height}}{\text{base}}$

Gradient $= \frac{6}{4} = \frac{3}{2}$ or 1.5

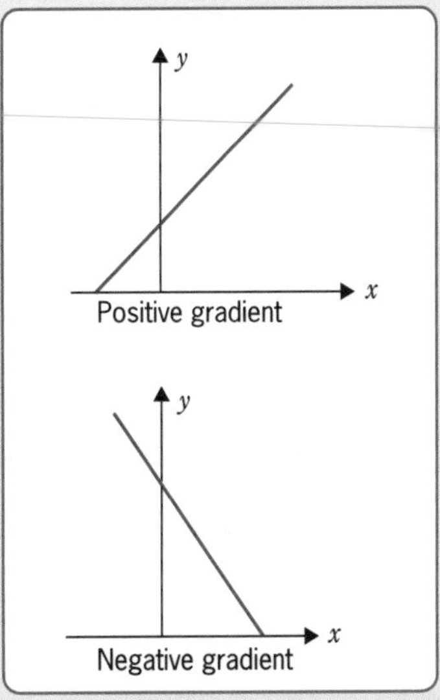

Positive gradient
Negative gradient

How are you feeling Grandad?
Look at my graph – I'm getting worse each day.
I know how to make you feel a bit better.
Your graph has slipped – you're fine !

CURVED GRAPHS

Draw the graph of $y = x^2 - 2x - 6$. Use values of x from –2 to 3.

STEP 1 Draw a table of values.

x	–2	–1	0	1	2	3
y	2	–3	–6	–7	–6	–3

STEP 2 Fill in the table of values by substituting the values of x into the equation.

i.e. $x = 1 \quad y = 1^2 - 2 \times 1 - 6,\ y = -7$

Coordinates are (1, –7)

STEP 3 Draw the axes on graph paper and plot the points.

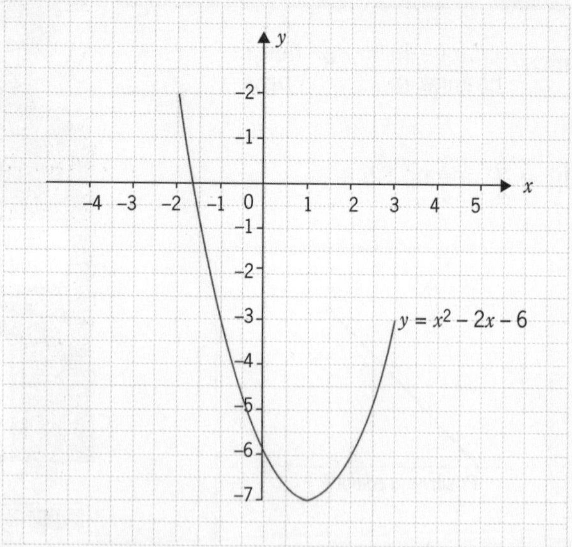

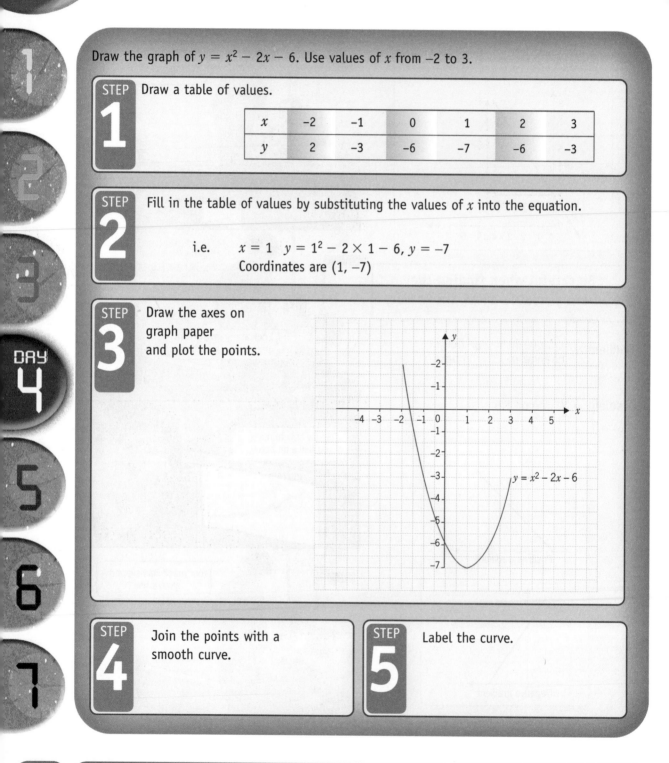

STEP 4 Join the points with a smooth curve.

STEP 5 Label the curve.

Quadratic graphs have an x^2 term.
They will be ∪ shaped if x^2 is positive, or ∩ shaped if x^2 is negative.

The minimum value is when $x = 1$, $y = -7$

The line of symmetry is at $x = 1$

Other graph shapes

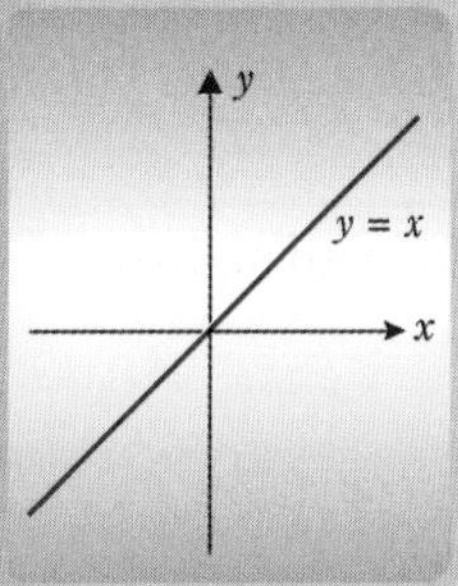

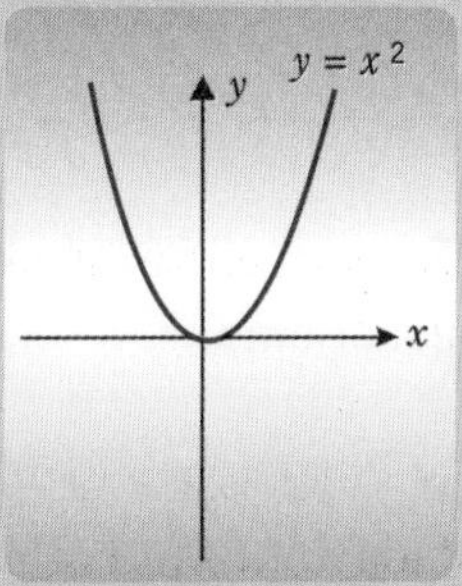

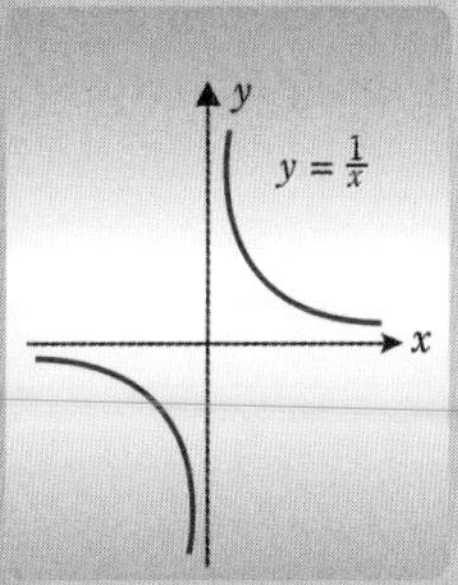

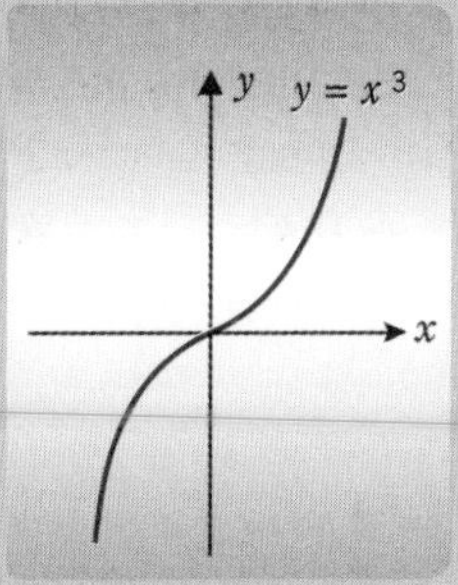

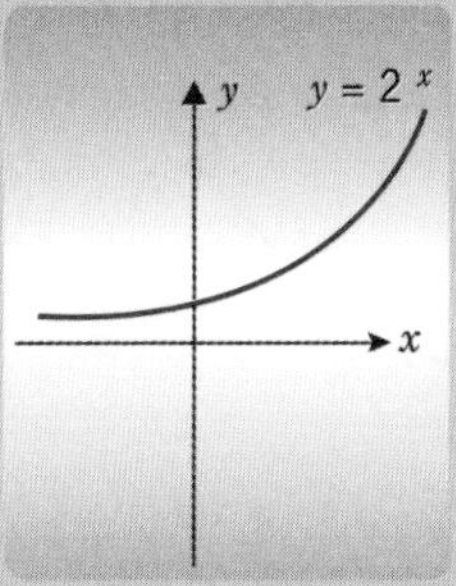

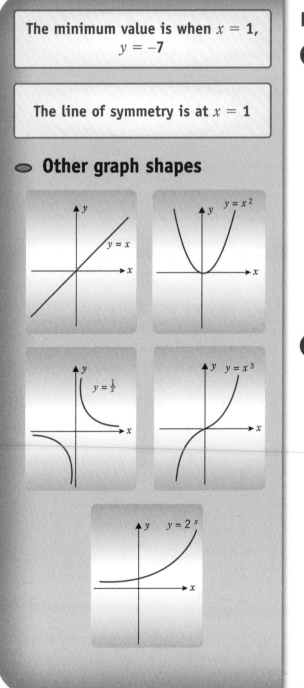

Progress check

1 a) Complete the table of values for $y = x^3 - 1$.

x	–3	–2	–1	0	1	2	3
y							

b) Draw the graph of $y = x^3 - 1$. Use scales of 2 units for 1 cm on the x-axis and 20 units for 1 cm on the y-axis.

c) From the graph find the value of x when $y = 15$.

2 Match each graph below to one of the equations.

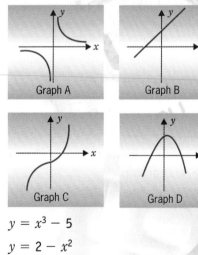

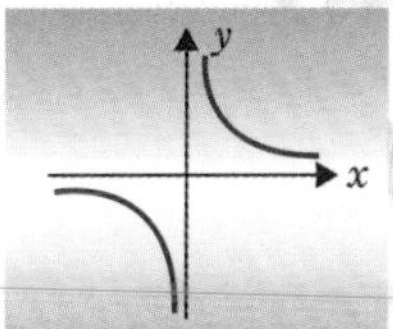

Graph A

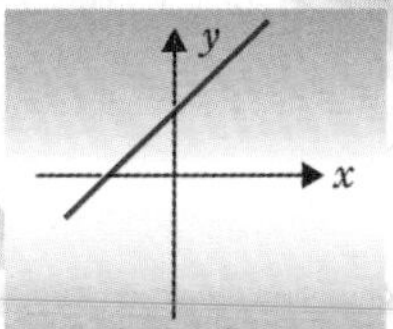

Graph B

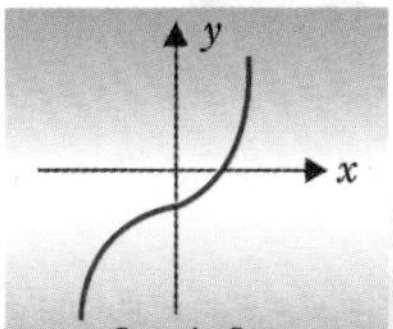

Graph C

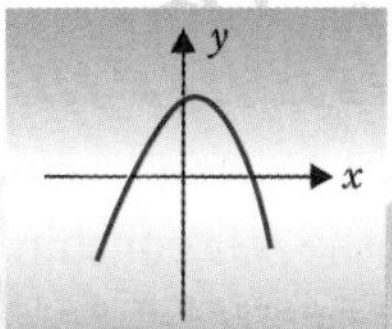

Graph D

$y = x^3 - 5$

$y = 2 - x^2$

$y = 4x + 2$

$y = \frac{3}{x}$

INTERPRETING GRAPHS

Using graphs to solve equations

Often an equation needs to be rearranged in order to resemble the equation of the plotted graph.

The graph $y = x^2 - 4x + 4$ is drawn here:

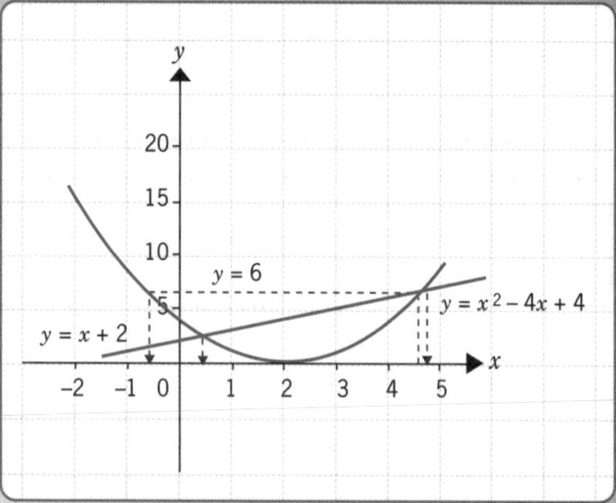

Use the graph to solve:

i) $x^2 - 4x + 4 = 6$
ii) $x^2 - 5x + 2 = 0$

For part (i) the equation needs to be arranged so it is the same as the plotted graph $y = x^2 - 4x + 4$

$x^2 - 4x + 4 = 6$

So the solution will be when the graph $y = x^2 - 4x + 4$ meets the line $y = 6$

$x = -0.4$ $x = 4.4$ (approximately)

For part (ii) we need to rearrange the equation:

$x^2 - 5x + 2 = 0$

$x^2 - 5x + 2 + x + 2 = x + 2$

(add $x + 2$ to both sides)

$x^2 - 4x + 4 = x + 2$

The solutions are where the graph crosses the line $y = x + 2$

$x = 0.4$ $x = 4.6$

Using graphs to find relationships

This graph is known to fit $y = pq^x$.

Use the graph to find the values of p and q and the relationship.

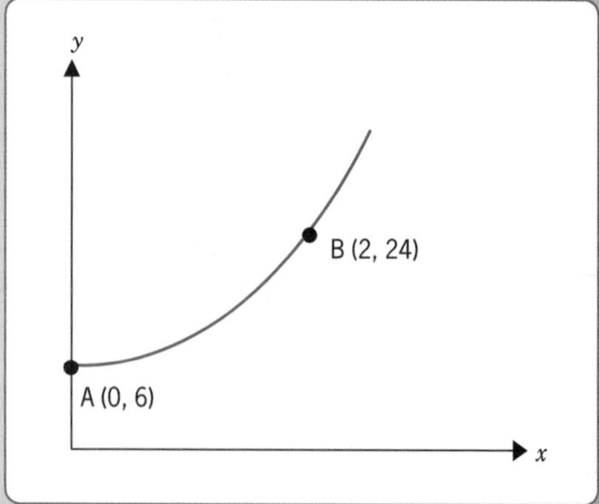

Accurate graphs can provide solutions to some equations.

20 MINS

STEP 1 Take the value at point A and substitute into $y = pq^x$

$6 = p \times q^0$

Since $q^0 = 1$ $\quad 6 = p$

STEP 2 Take the coordinates at point B and substitute into $y = pq^x$

$24 = 6 \times q^2$

$\therefore \quad 4 = q^2$

$q = 2$

The relationship is $y = 6 \times 2^x$

Speed–time graphs

- Distance is the area between the graph and x-axis.
- A positive gradient means the speed is increasing.

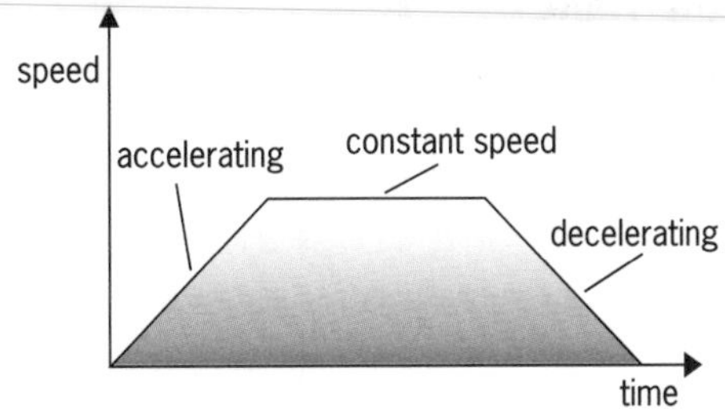

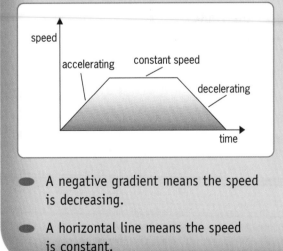

- A negative gradient means the speed is decreasing.
- A horizontal line means the speed is constant.

Progress check

Draw the graph of $y = x^2 + x - 2$.
Use your graph to decide whether the solutions of these equations are true or false.

1. $x^2 + x = 5$, solution is approximately $x = -2.8$, $x = 1.8$
2. $x^2 - 2 = 0$, solution is approximately $x = 2.4$, $x = -2.4$
3. $x^2 - 2x - 2 = 0$, solution is approximately $x = -0.7$, $x = 5.6$

1 2 3 DAY 4 5 6 7

FUNCTIONS AND TRANSFORMATIONS

$y = f(x) \pm a$

This is when the graphs move up or down the y-axis by a value of a.

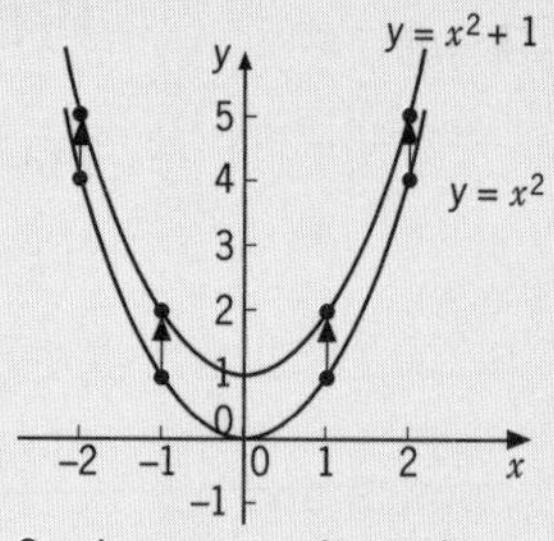

Graph moves up the y-axis by one unit.

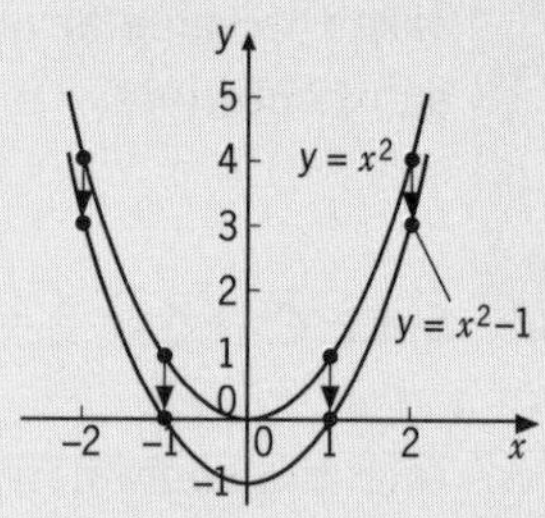

Graph moves down the y-axis by one unit.

$y = f(x \pm a)$

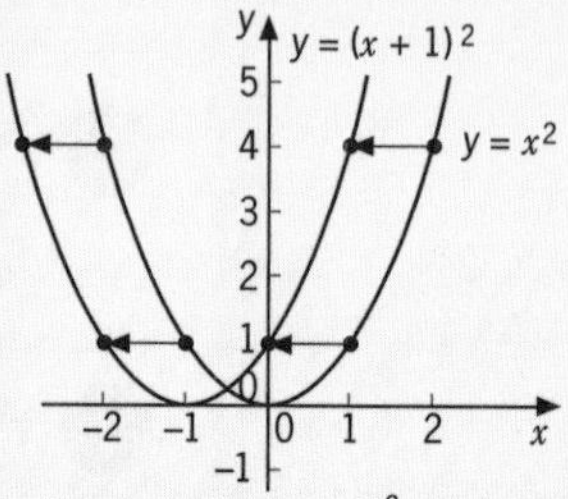

Graph of $y = (x + 1)^2$ moves one unit to the left.

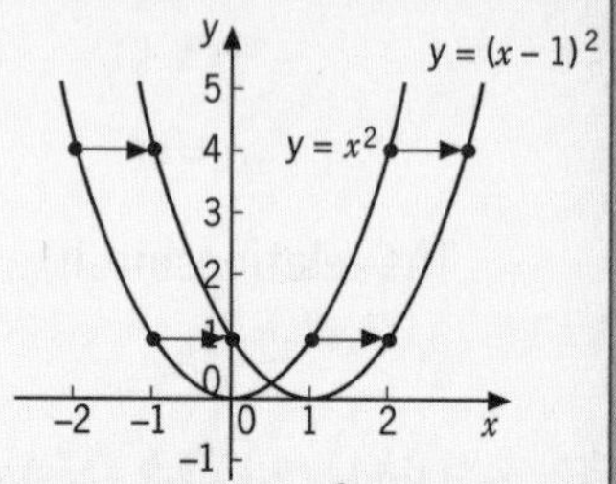

Graph of $y = (x - 1)^2$ moves to the right one unit.

This is when the graphs move along the x-axis by a units.

$y = f(x + a)$ moves the graph a units to the left

$y = f(x - a)$ moves the graph a units to the right

$y = kf(x)$

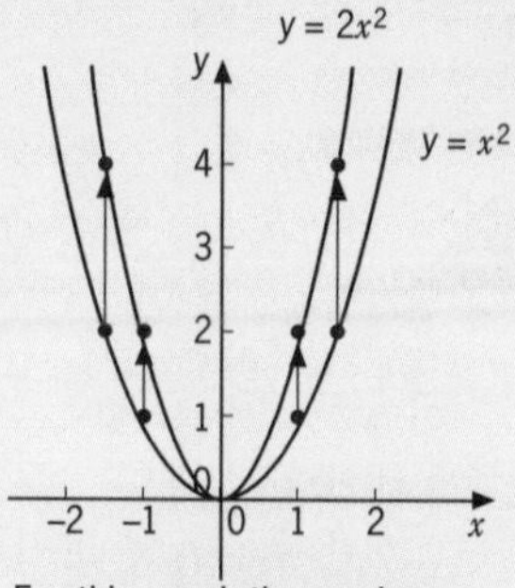

For this graph the x values stay the same and the y values are multiplied by 2.

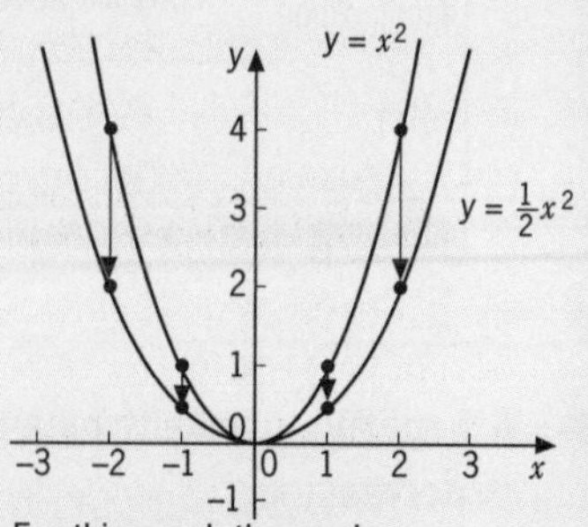

For this graph the x values stay the same and the y values are multiplied by $\frac{1}{2}$.

If y = an expression involving x then it can be written as $y = f(x)$. The graphs of the related functions can be found by applying transformations.

15 MINS

- **This is when the original graph stretches along the y-axis by a factor of k.**
- **If $k > 1$ then the points are stretched upwards in the y direction of a scale factor of k.**
- **If $k < 1$, e.g. $y = \frac{1}{2}x^2$, the graph squashes downwards by a scale factor of $\frac{1}{2}$.**

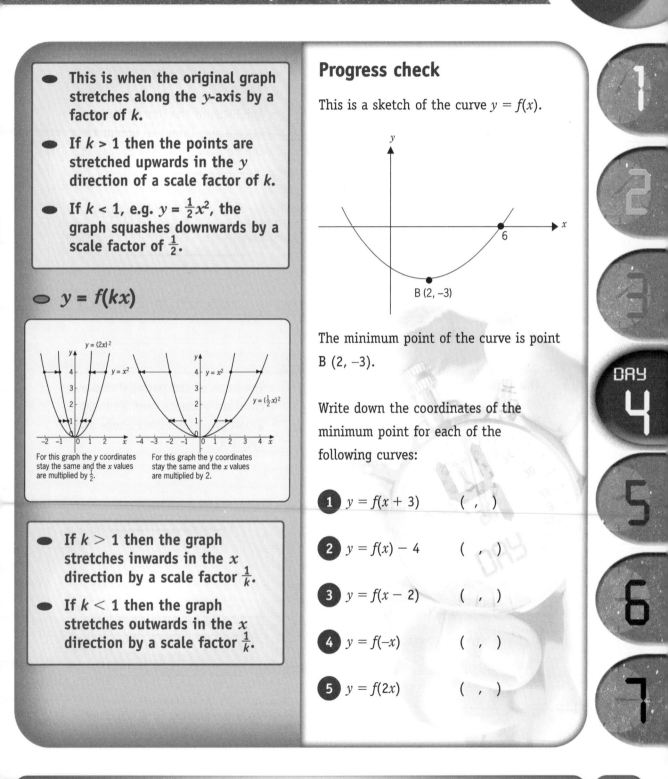

$y = f(kx)$

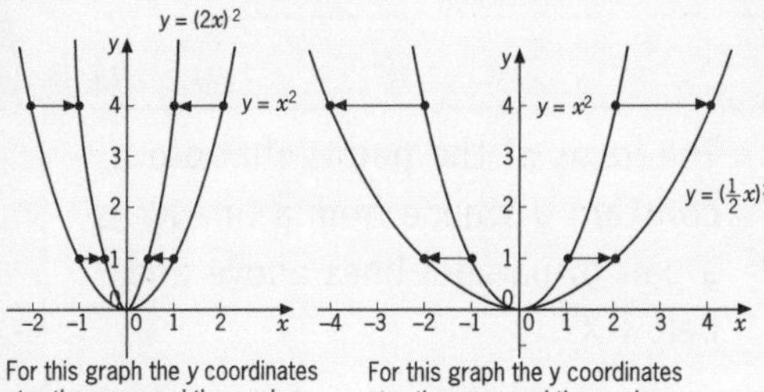

For this graph the y coordinates stay the same and the x values are multiplied by $\frac{1}{2}$.

For this graph the y coordinates stay the same and the x values are multiplied by 2.

- **If $k > 1$ then the graph stretches inwards in the x direction by a scale factor $\frac{1}{k}$.**
- **If $k < 1$ then the graph stretches outwards in the x direction by a scale factor $\frac{1}{k}$.**

Progress check

This is a sketch of the curve $y = f(x)$.

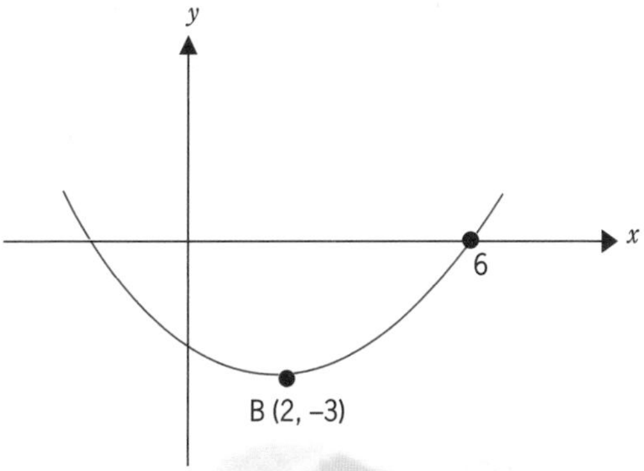

The minimum point of the curve is point B (2, –3).

Write down the coordinates of the minimum point for each of the following curves:

1. $y = f(x + 3)$ (,)
2. $y = f(x) - 4$ (,)
3. $y = f(x - 2)$ (,)
4. $y = f(-x)$ (,)
5. $y = f(2x)$ (,)

LOCI

Types of loci

1. The locus of the points that are a constant distance from a fixed point is a circle.

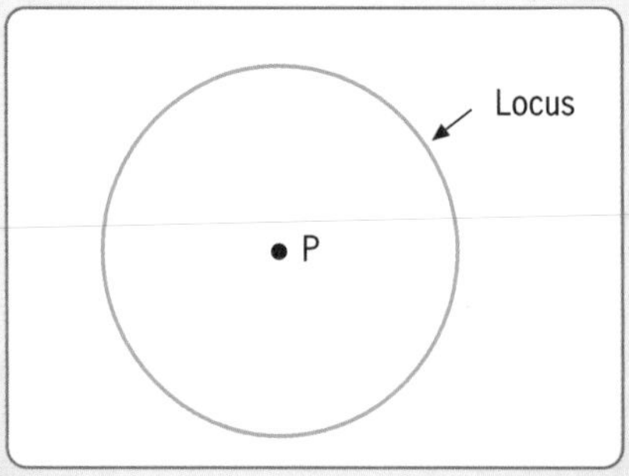

2. The locus of the points that are equidistant from two points XY is the perpendicular bisector of XY.

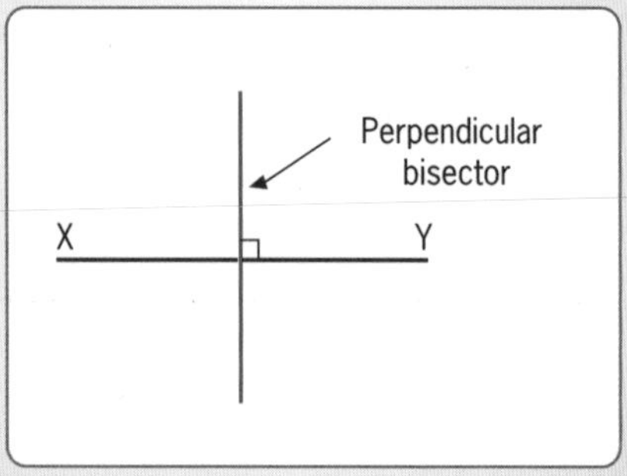

3. The locus of the points that are equidistant from two lines is the line that bisects the angle between the lines.

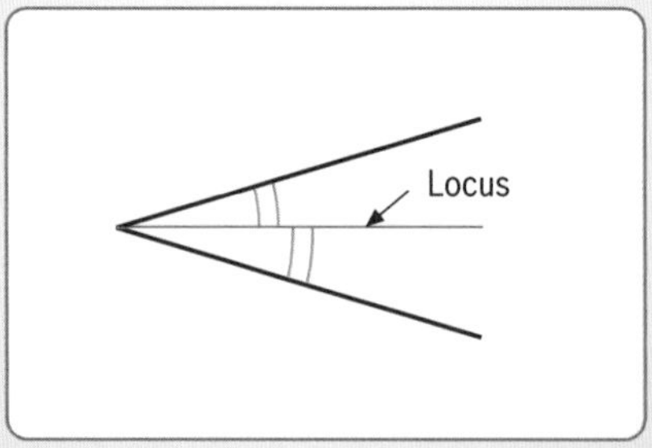

4. The locus of the points that are a constant distance from a line XY is a pair of parallel lines above and below XY.

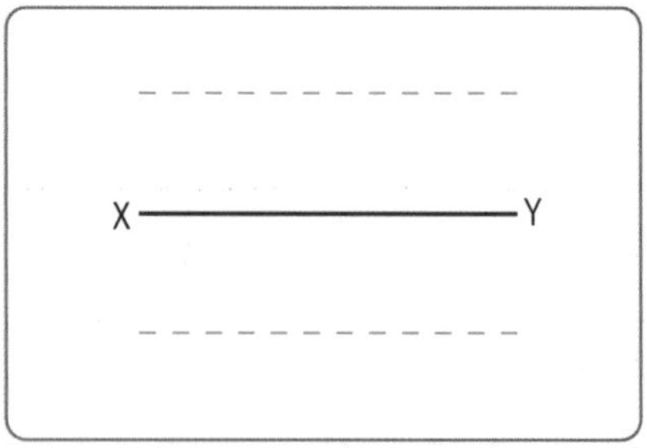

Sometimes you need to combine types 1 and 3.

A fixed distance from a line segment gives this locus.

The locus of a point is the set of all the possible positions that the point can occupy subject to some given condition or rule.

10 MINS

Example

Three radio transmitters form an equilateral triangle ABC with sides of 50 km. The range of the transmitter at A is 37.5 km, at B 30 km and at C 28 km. Using a scale of 1 cm to 10 km, construct a scale diagram to show where signals from all three transmitters can be received.

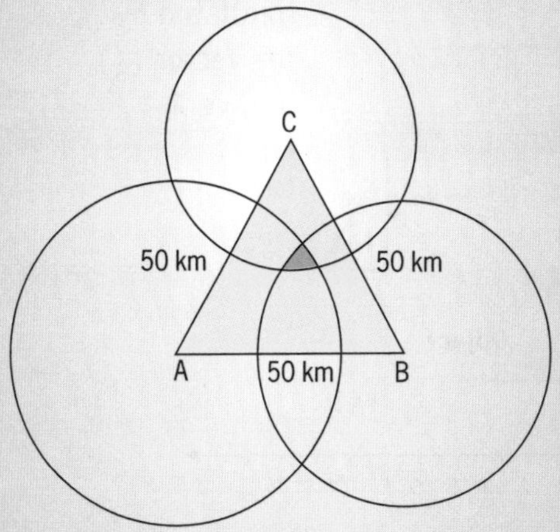

Please note that on your scale drawing the circle at A would have a radius of 3.75 cm. The circle at B would have a radius of 3 cm and the circle at C a radius of 2.8 cm.

Progress check

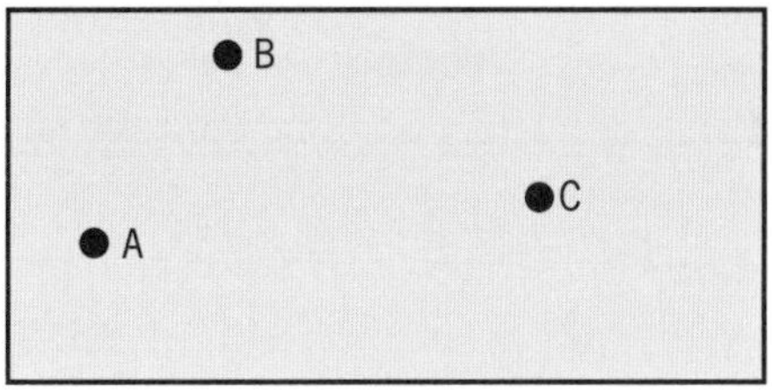

The plan shows a garden drawn to a scale of 1 cm: 2 m A and B are bushes and C is a pond. A landscape gardener has decided:

a) to lay a path right across the garden at an equal distance from each of the bushes.

b) to lay a flower border 4 m wide around pond C.

Construct these features on the plan above.

DAY 4

TRANSFORMATIONS

Translations

Translations move figures from one position to another position. **Vectors** are used to describe the distance and direction of the translations.

A vector is written as $\begin{pmatrix} a \\ b \end{pmatrix}$

a represents the horizontal distance

b represents the vertical distance

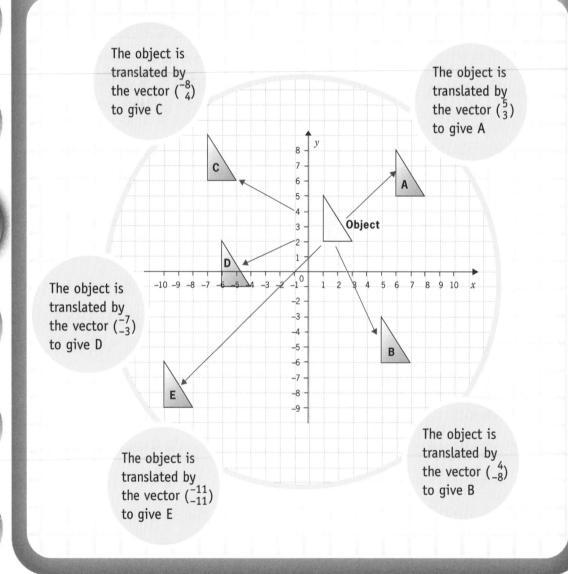

The object is translated by the vector $\begin{pmatrix} -8 \\ 4 \end{pmatrix}$ to give C

The object is translated by the vector $\begin{pmatrix} 5 \\ 3 \end{pmatrix}$ to give A

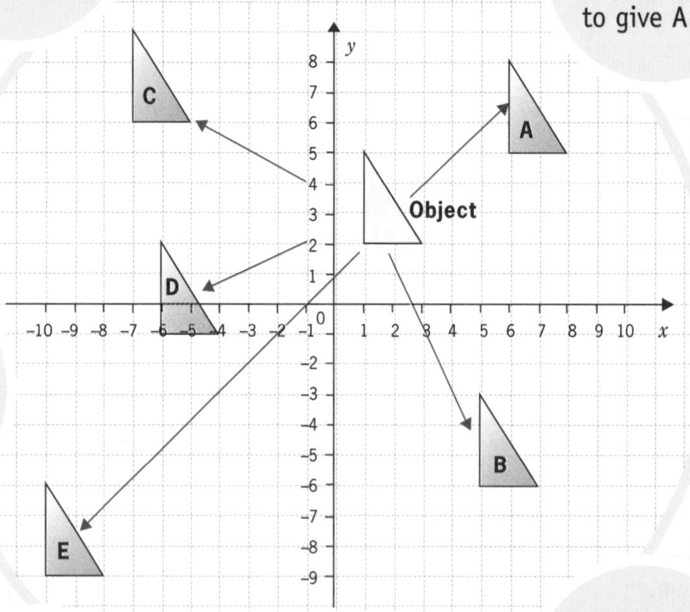

The object is translated by the vector $\begin{pmatrix} -7 \\ -3 \end{pmatrix}$ to give D

The object is translated by the vector $\begin{pmatrix} -11 \\ -11 \end{pmatrix}$ to give E

The object is translated by the vector $\begin{pmatrix} 4 \\ -8 \end{pmatrix}$ to give B

A transformation changes the position or size of a shape. There are four types of transformation: reflection, rotation, translation and enlargement.

> **The object and the image are congruent when the shape is translated.**

Rotations

In a rotation the object is turned by a given angle about a fixed point called the **centre of rotation**. The size and shape of the figure are not changed.

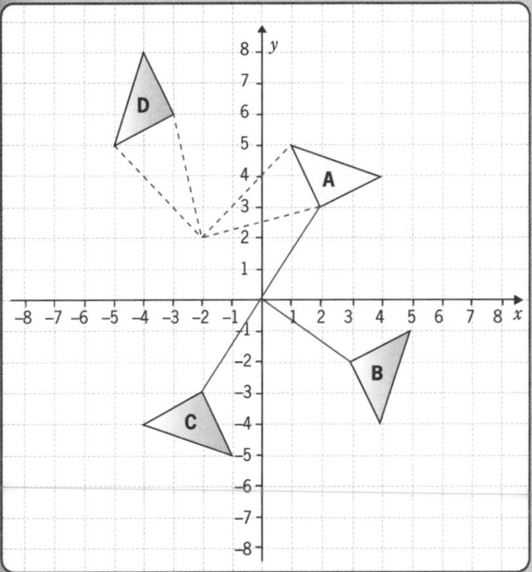

The object A is rotated 90° anticlockwise about (–2, 2) to give the image D.

Object A is rotated by 180° about (0, 0) to give the image C.

The object A is rotated by 90° clockwise about (0, 0) to give the image B.

Please send me a photo
a little bit of reflection...
mirror, mirror on the wall...

ENLARGEMENTS

Examples

Describe fully the transformation that maps ABC onto A'B'C'.

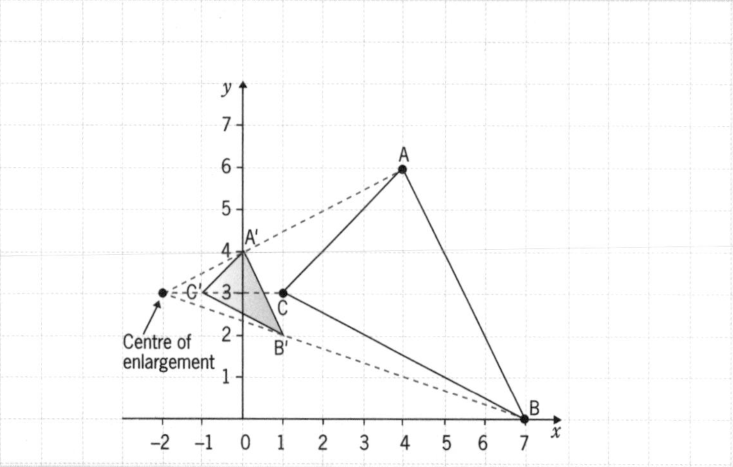

- To find the centre of enlargement join A to A', B to B' etc. and continue the line.
- Where all the lines meet is the centre of enlargement: (−2, 3).
- The transformation is an enlargement of scale factor $\frac{1}{3}$. Centre of enlargement is (−2, 3).

Enlargements change the size but not the shape of the object. The centre of enlargement is the point from which the enlargement takes place.

15 MINS

Enlargements with a negative scale factor

For an enlargement with a negative scale factor, the image is situated on the opposite side of the centre of enlargement.

The triangle A (3, 4), B (9, 4) and C (3, 10) is enlarged with a scale factor of $^{-}\frac{1}{3}$, about the centre of enlargement (0, 1). Label the enlargement A′B′C′.

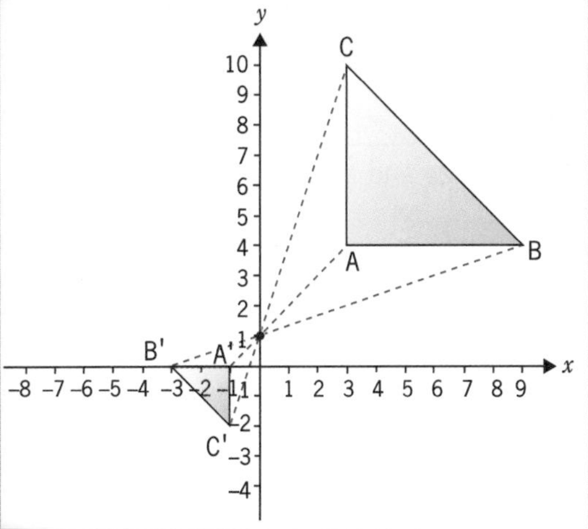

- Notice A′B′C′ is a third of the size of triangle ABC.
- ABC is on the opposite side of the centre of enlargement.

Progress check

1. Complete the diagram below, to show the enlargement of the shape by a scale factor of $\frac{1}{2}$. Centre of enlargement at (0, 0). Call the shape T.

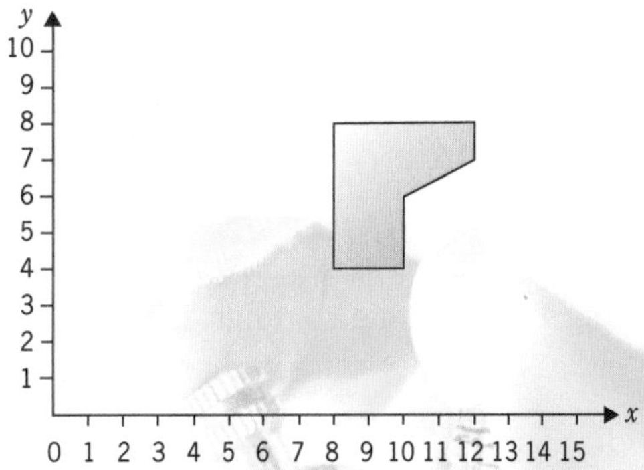

2. The quadrilateral is enlarged with a scale factor of $^{-}\frac{1}{2}$ about the origin. Draw the enlargement on the diagram below.

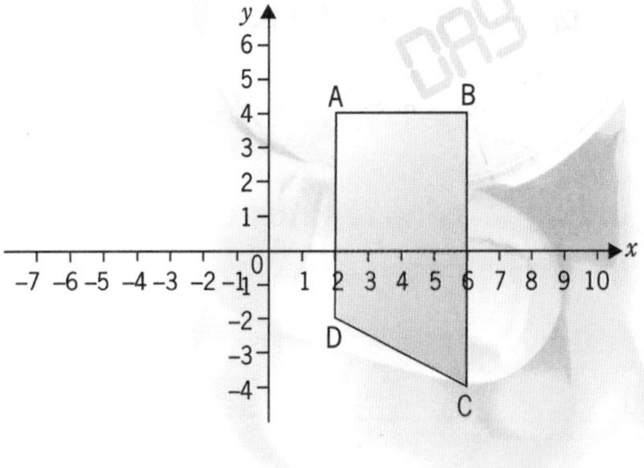

DAY 4

SIMILARITY AND CONGRUENCY 1

Congruent triangles

Two triangles are congruent if one of the following sets of conditions is true (S = side, A = angle, R = right angle, H = hypotenuse).

SSS: The three pairs of sides are equal.

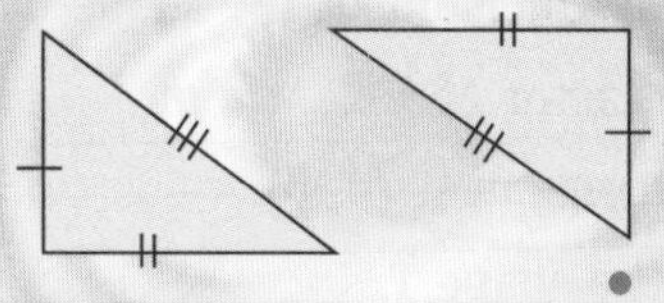

SAS: Two pairs of sides and the included angle are equal.

RHS: Each triangle contains a right angle. The hypotenuse and another pair of sides are equal.

AAS: Two angles and a corresponding side are equal.

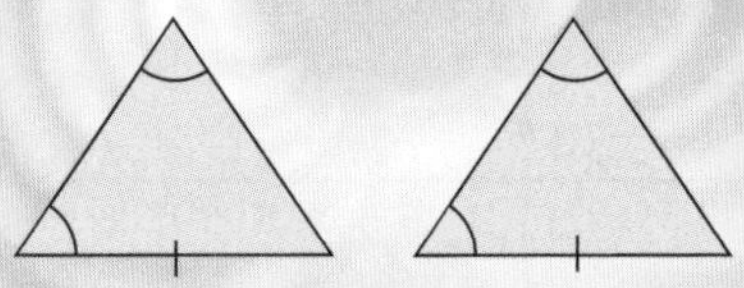

Similarity

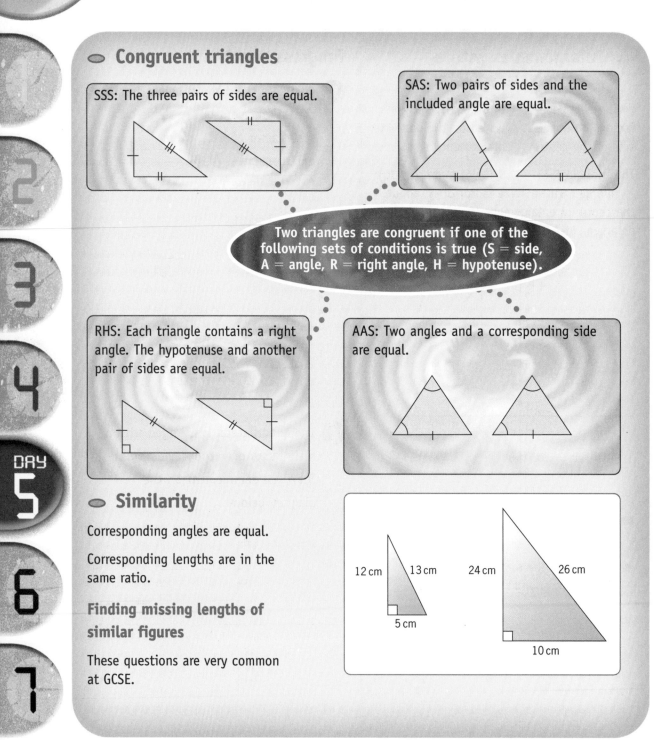

Corresponding angles are equal.

Corresponding lengths are in the same ratio.

Finding missing lengths of similar figures

These questions are very common at GCSE.

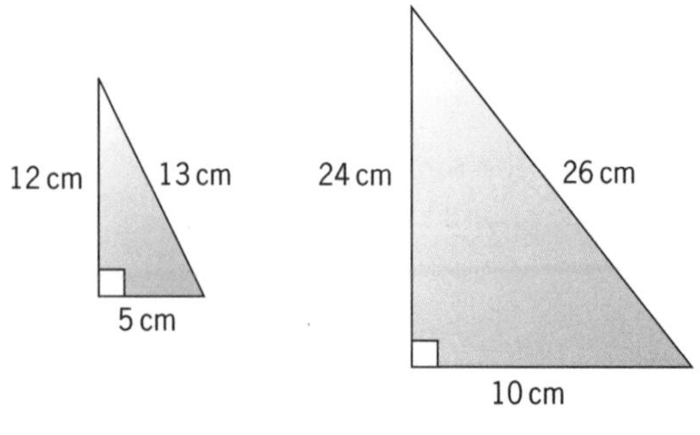

Objects that are exactly the same shape but different sizes are called similar shapes. One is an enlargement of the other.

Examples

Find the missing lengths labelled *a* in the diagrams below:

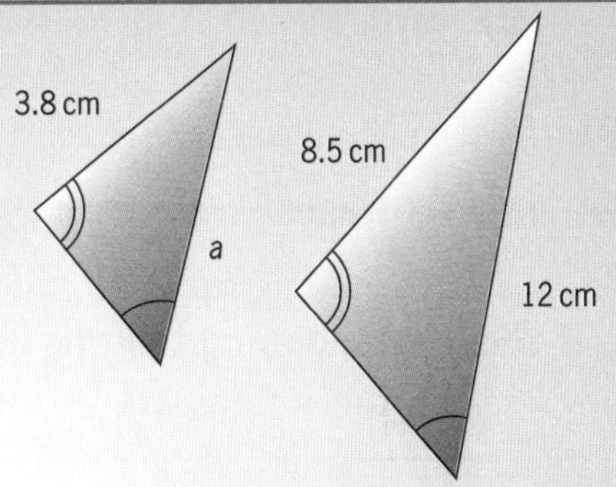

a) $\frac{a}{12} = \frac{3.8}{8.5}$

- Corresponding lengths are in the same ratio.

$a = \frac{3.8}{8.5} \times 12$ — Multiply both sides by 12.

$a = 5.36$ cm (3sf)

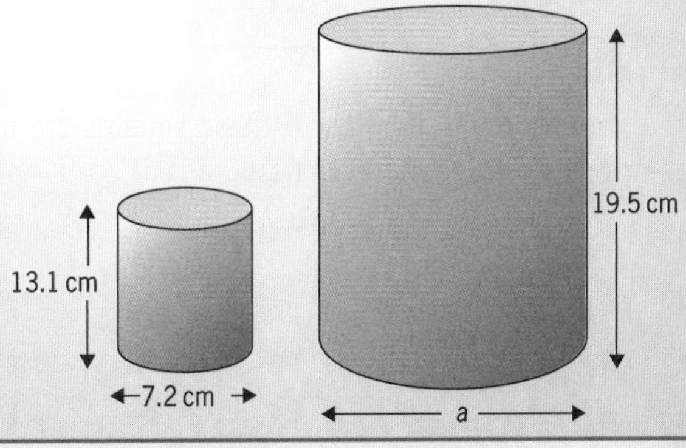

b) $\frac{a}{7.2} = \frac{19.5}{13.1}$

$a = \frac{19.5}{13.1} \times 7.2$ — Multiply both sides by 7.2.

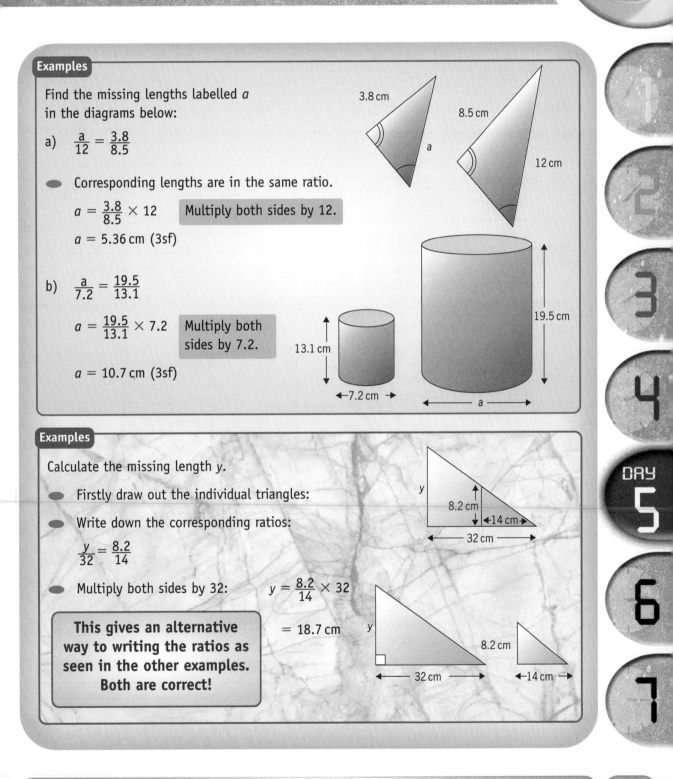

$a = 10.7$ cm (3sf)

Examples

Calculate the missing length *y*.

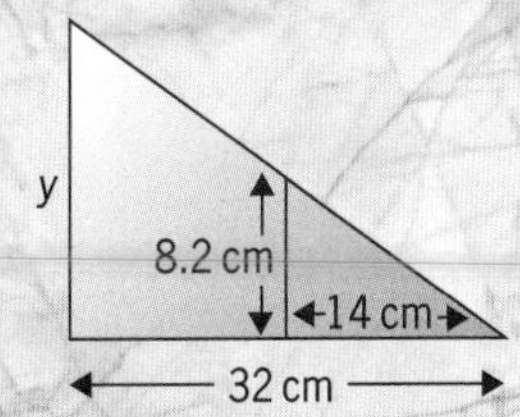

- Firstly draw out the individual triangles:
- Write down the corresponding ratios:

$\frac{y}{32} = \frac{8.2}{14}$

- Multiply both sides by 32: $y = \frac{8.2}{14} \times 32$

$= 18.7$ cm

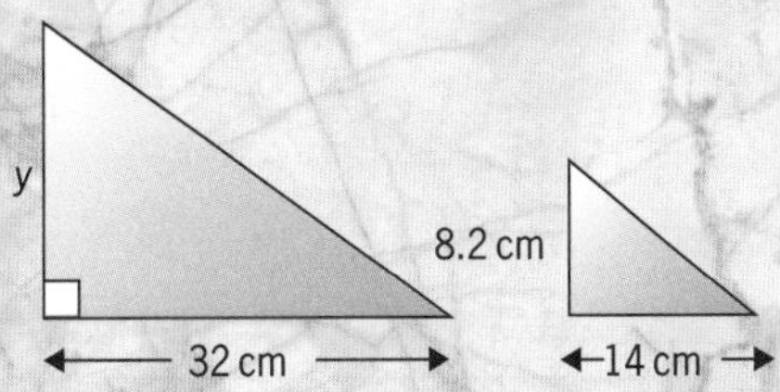

This gives an alternative way to writing the ratios as seen in the other examples. Both are correct!

SIMILARITY AND CONGRUENCY 2

Area and volume of similar figures

Areas

Areas of similar figures are not in the same ratio as their lengths.

If the corresponding lengths are in the ratio $a : b$ then their areas are in the ratio $a^2 : b^2$.

Examples

The corresponding lengths of these squares are in the ratio 1 : 3 and their areas are in the ratio 1 : 9.

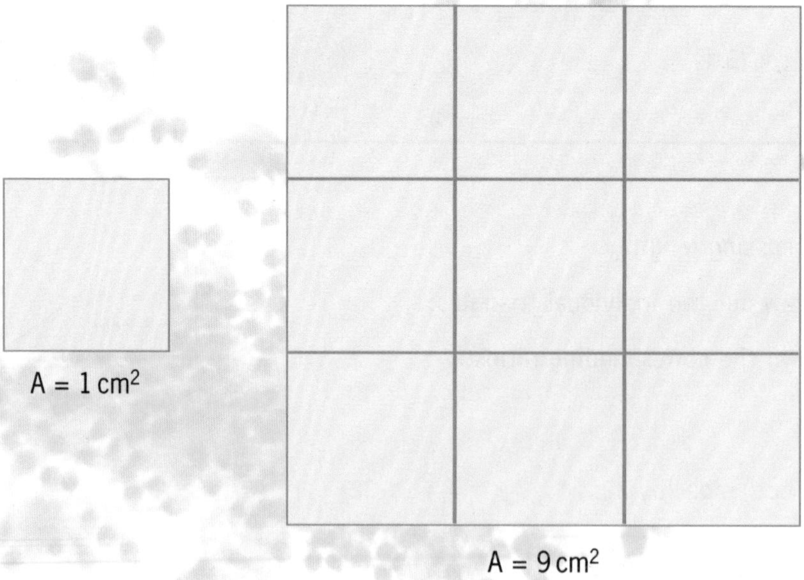

Volumes

A similar result can be found when looking at the volumes of similar figures.

If the corresponding lengths are in the ratio $a : b$, then their volumes are in the ratio $a^3 : b^3$.

For a scale factor n:
The sides are n times bigger
The areas are n^2 times bigger
The volumes are n^3 times bigger

Example

These two solids are similar. If the volume of the smaller solid is 9 cm³, calculate the volume of the larger solid.

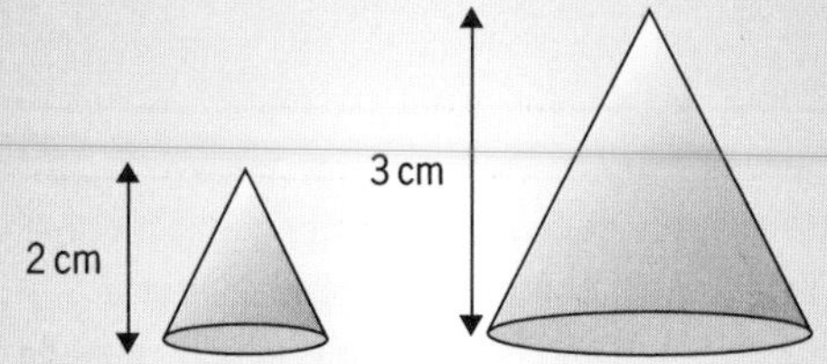

Linear scale factor is 2 : 3

Volume scale factor is $2^3 : 3^3$

Volume of larger solid is $\frac{27}{8} \times 9$

$= 30.375\ \text{cm}^3$

Progress check

1. Calculate the lengths marked n in these similar shapes. Give your answers correct to 1dp.

a)

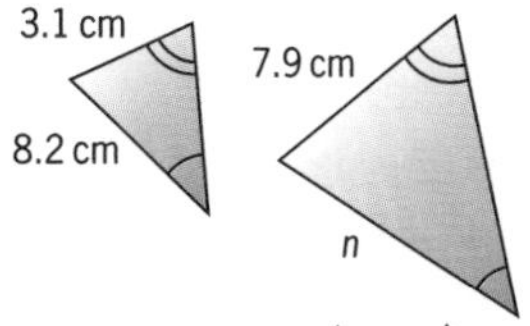

b)

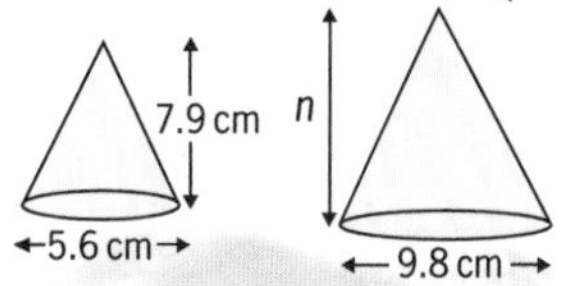

c)

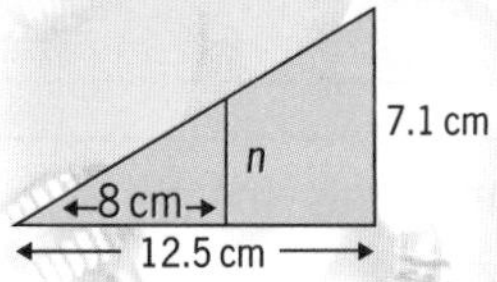

2. The heights of two similar shapes are 8 cm and 12 cm. If the area of the larger shape is 64 cm², find the area of the smaller shape.

3. Are these two triangles congruent? Explain why.

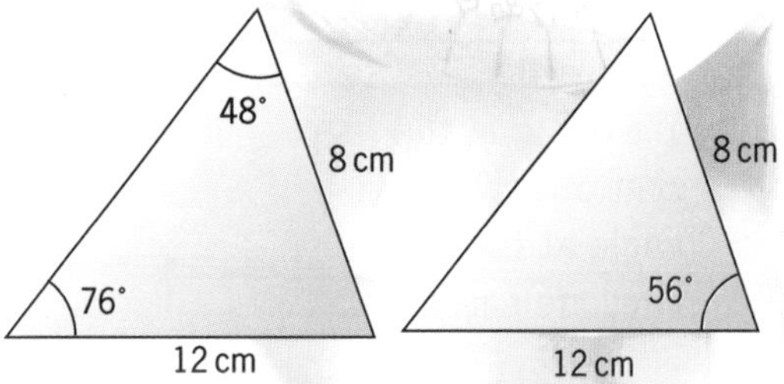

CIRCLE THEOREMS

The circle theorems

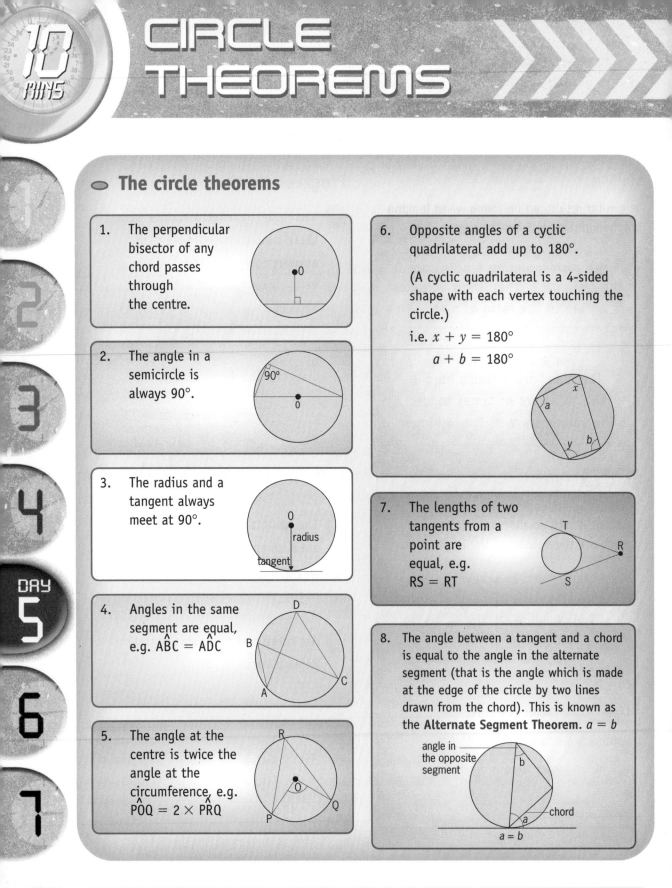

1. The perpendicular bisector of any chord passes through the centre.

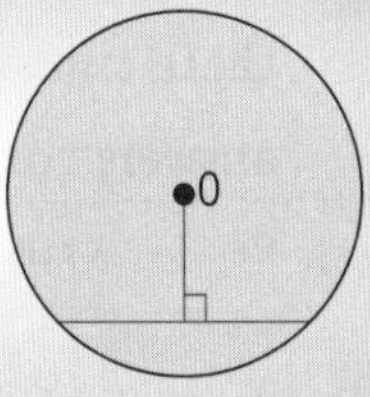

2. The angle in a semicircle is always 90°.

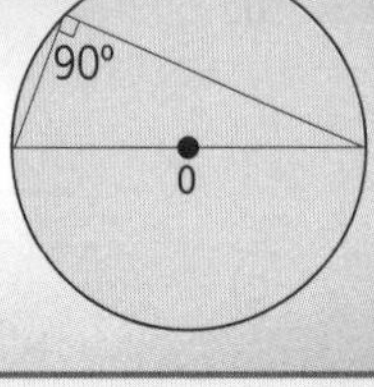

3. The radius and a tangent always meet at 90°.

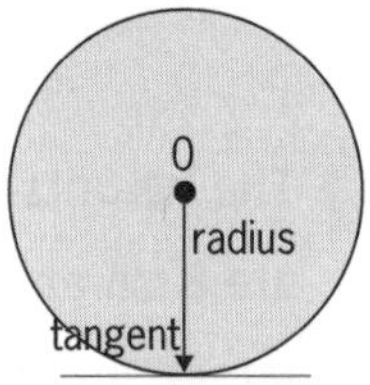

4. Angles in the same segment are equal, e.g. $A\hat{B}C = A\hat{D}C$

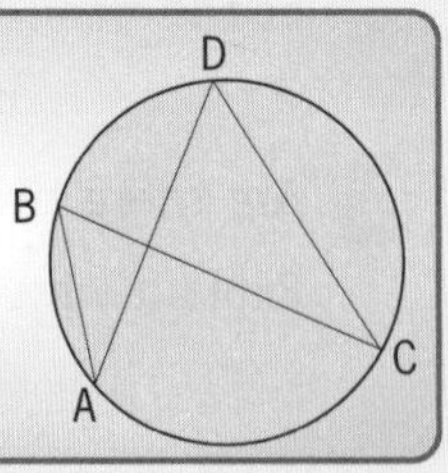

5. The angle at the centre is twice the angle at the circumference, e.g. $P\hat{O}Q = 2 \times P\hat{R}Q$

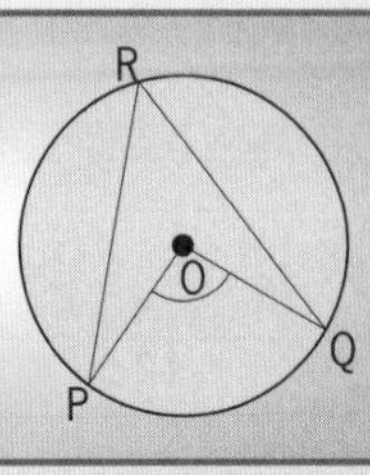

6. Opposite angles of a cyclic quadrilateral add up to 180°.

(A cyclic quadrilateral is a 4-sided shape with each vertex touching the circle.)

i.e. $x + y = 180°$

$a + b = 180°$

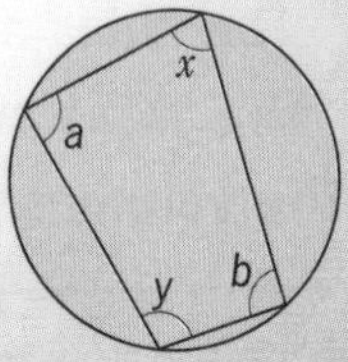

7. The lengths of two tangents from a point are equal, e.g. RS = RT

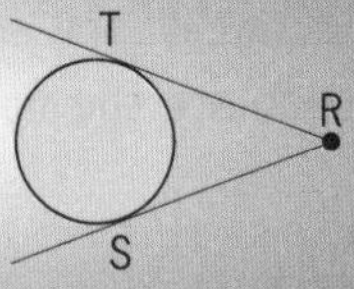

8. The angle between a tangent and a chord is equal to the angle in the alternate segment (that is the angle which is made at the edge of the circle by two lines drawn from the chord). This is known as the **Alternate Segment Theorem.** $a = b$

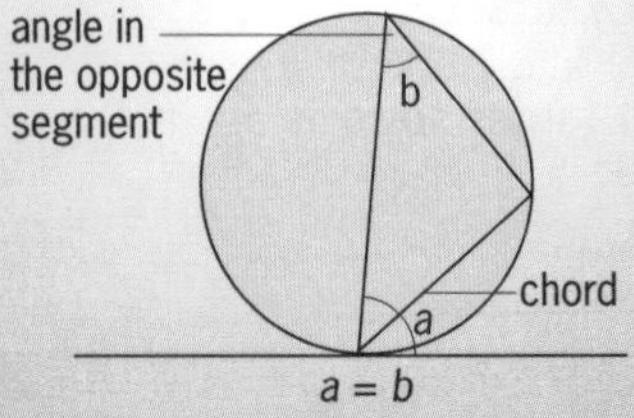

There are several theorems about circles that you need to know.

Example

Calculate the angles marked *a*–*d* in the diagram below. Give a reason for your answers.

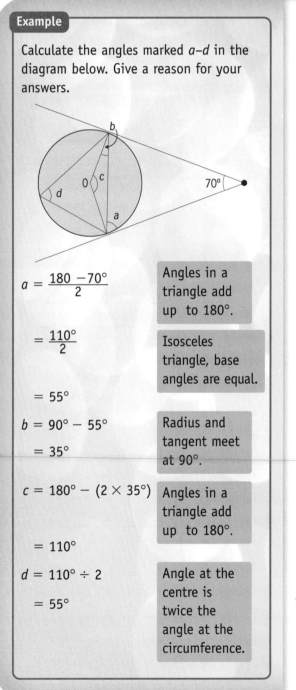

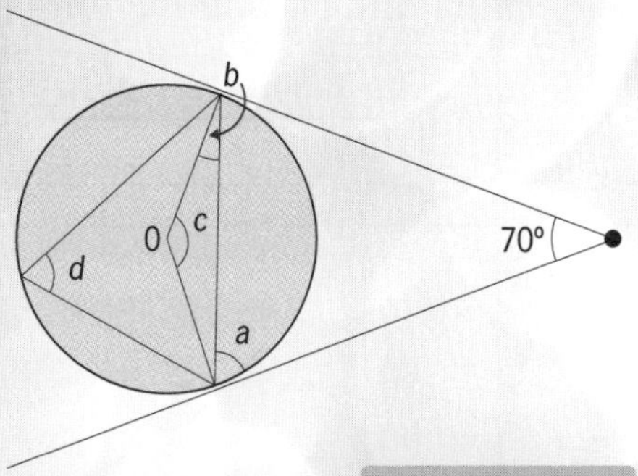

$a = \frac{180 - 70°}{2}$ — Angles in a triangle add up to 180°.

$= \frac{110°}{2}$ — Isosceles triangle, base angles are equal.

$= 55°$

$b = 90° - 55°$ — Radius and tangent meet at 90°.

$= 35°$

$c = 180° - (2 \times 35°)$ — Angles in a triangle add up to 180°.

$= 110°$

$d = 110° \div 2$ — Angle at the centre is twice the angle at the circumference.

$= 55°$

Progress check

Some angles are written on cards. Match the missing angles in the diagrams below with the correct card. 0 represents the centre of the circle.

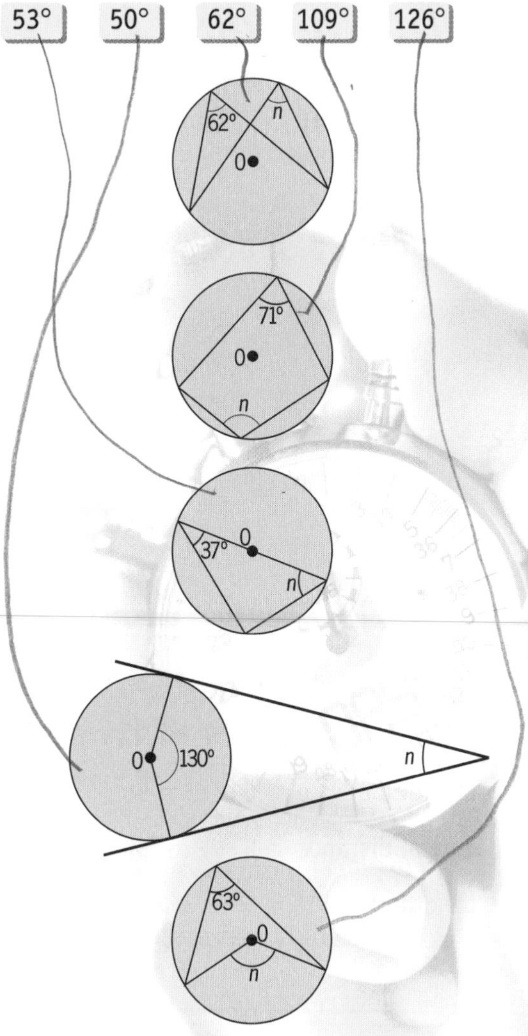

1
2
3
4
DAY 5
6
7

PYTHAGORAS' THEOREM

$a^2 + b^2 = c^2$

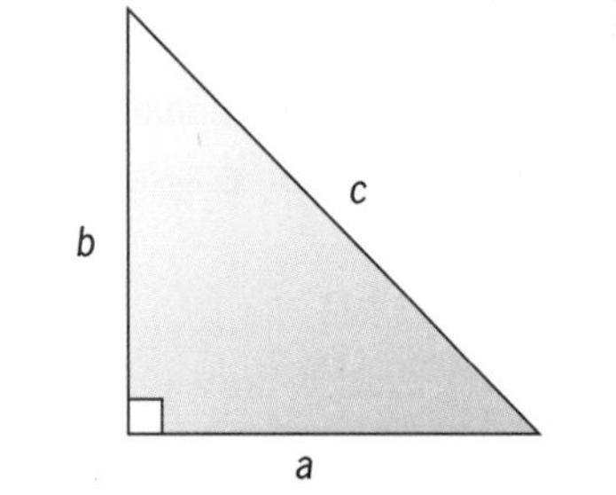

Finding the hypotenuse (n)

$n^2 = 7^2 + 12^2$ Square the two sides.

$n^2 = 49 + 144$ Add the two sides together.

$n^2 = 193$

$n = \sqrt{193}$ Square root.

$n = 13.9$ cm (3sf) Round to 3sf.

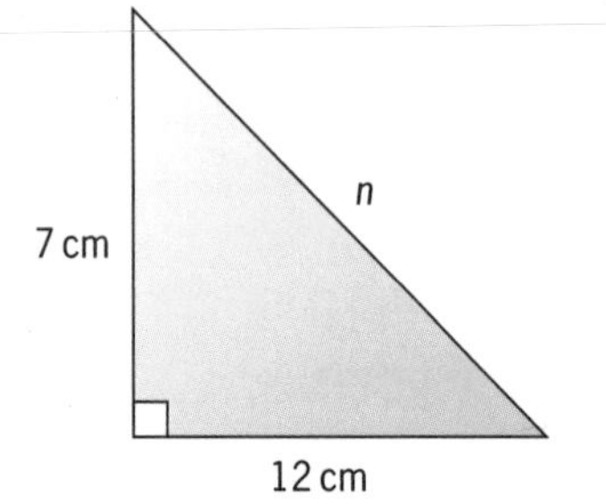

Finding a short side (p)

$15^2 = p^2 + 8^2$

$15^2 - 8^2 = p^2$ When finding a shorter length remember to subtract.

$225 - 64 = p^2$

$161 = p^2$

$\sqrt{161} = p$

$p = 12.7$ cm (3 sf)

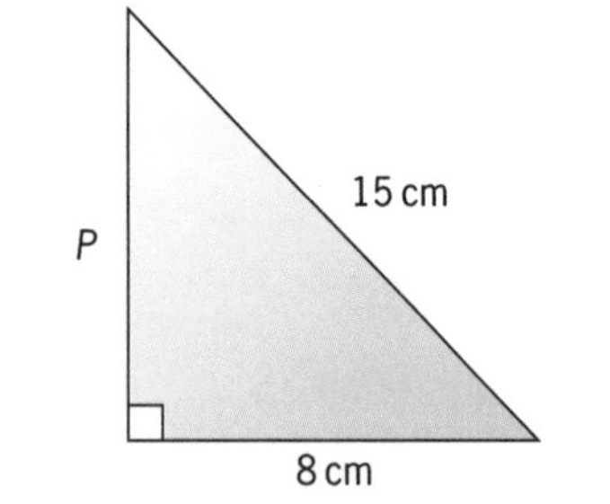

Finding the length of a line AB, given the coordinates of its end points

Horizontal distance $= 6 - 1 = 5$

Vertical distance $= 5 - 2 = 3$

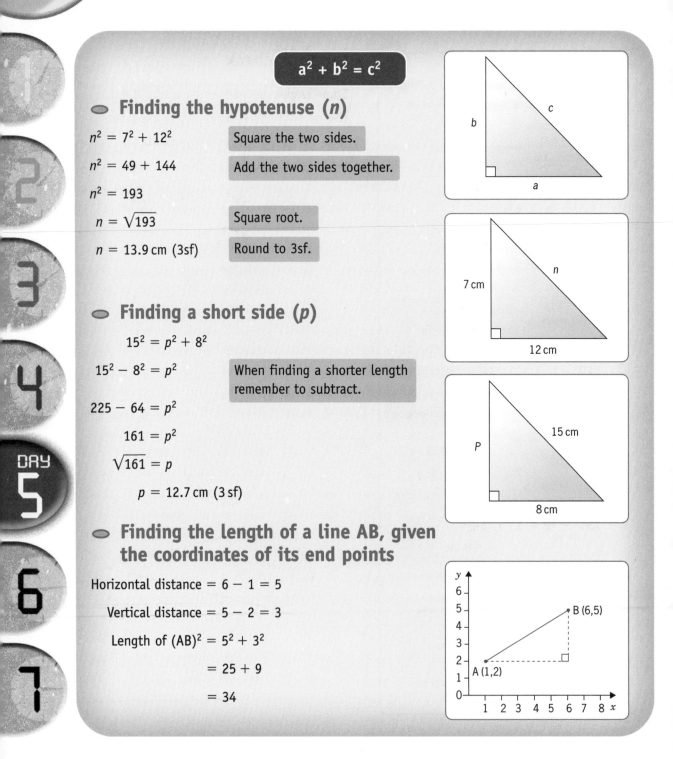

Length of $(AB)^2 = 5^2 + 3^2$

$= 25 + 9$

$= 34$

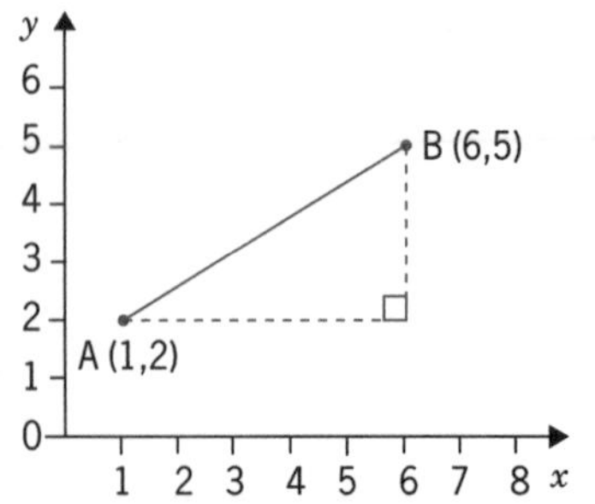

The hypotenuse is the longest side of a right-angled triangle. It is always opposite the right angle.

$AB = \sqrt{34}$

$AB = 5.83$ cm

We could leave this as $\sqrt{34}$. This is known as leaving in surd form.

Solving a more difficult problem

Calculate the vertical height of this isosceles triangle.

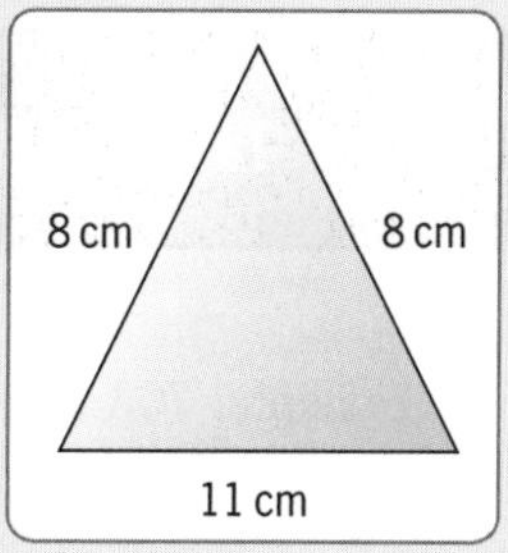

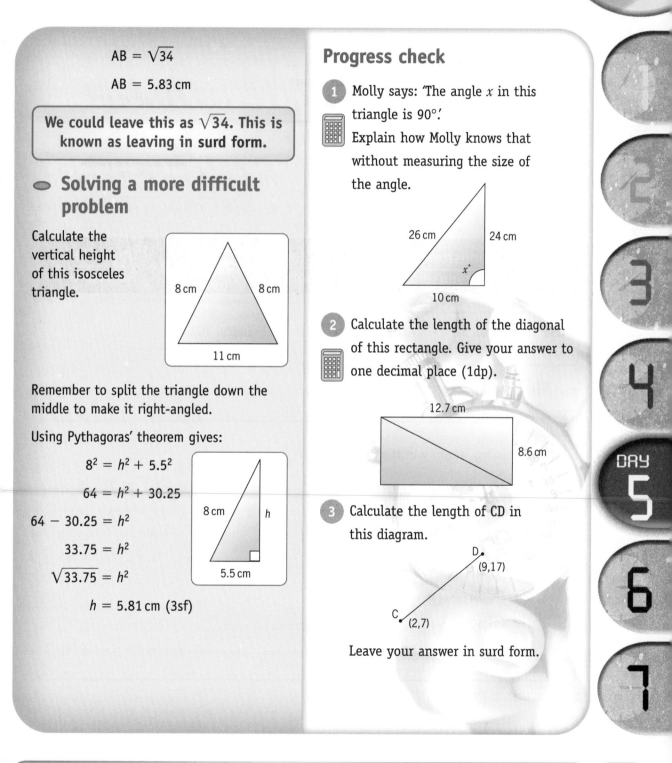

Remember to split the triangle down the middle to make it right-angled.

Using Pythagoras' theorem gives:

$8^2 = h^2 + 5.5^2$

$64 = h^2 + 30.25$

$64 - 30.25 = h^2$

$33.75 = h^2$

$\sqrt{33.75} = h^2$

$h = 5.81$ cm (3sf)

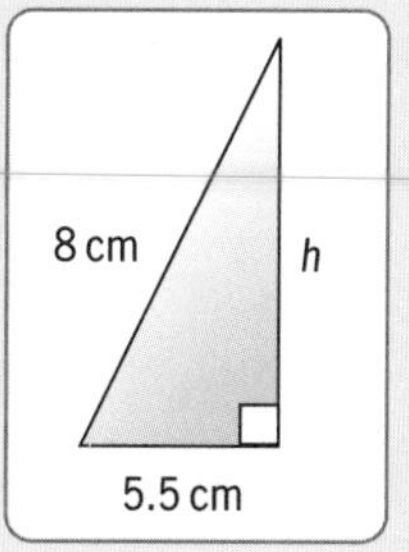

Progress check

1. Molly says: 'The angle x in this triangle is 90°.' Explain how Molly knows that without measuring the size of the angle.

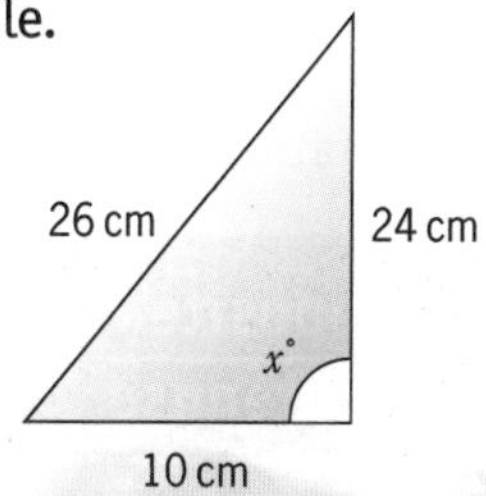

2. Calculate the length of the diagonal of this rectangle. Give your answer to one decimal place (1dp).

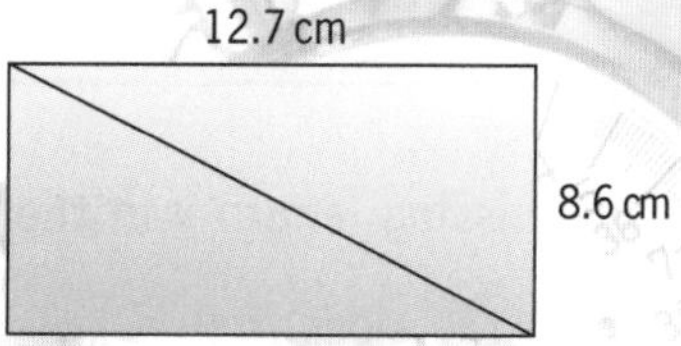

3. Calculate the length of CD in this diagram.

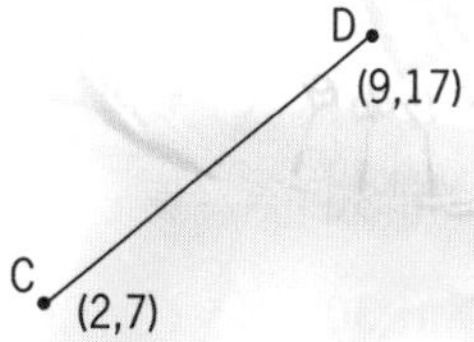

Leave your answer in surd form.

TRIGONOMETRY IN RIGHT-ANGLED TRIANGLES

15 MINS

The sides of a right-angled triangle are given temporary names according to where they are in relation to a chosen angle θ.

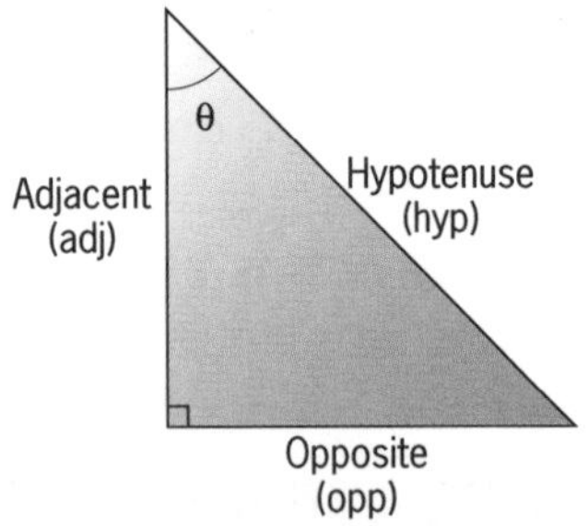

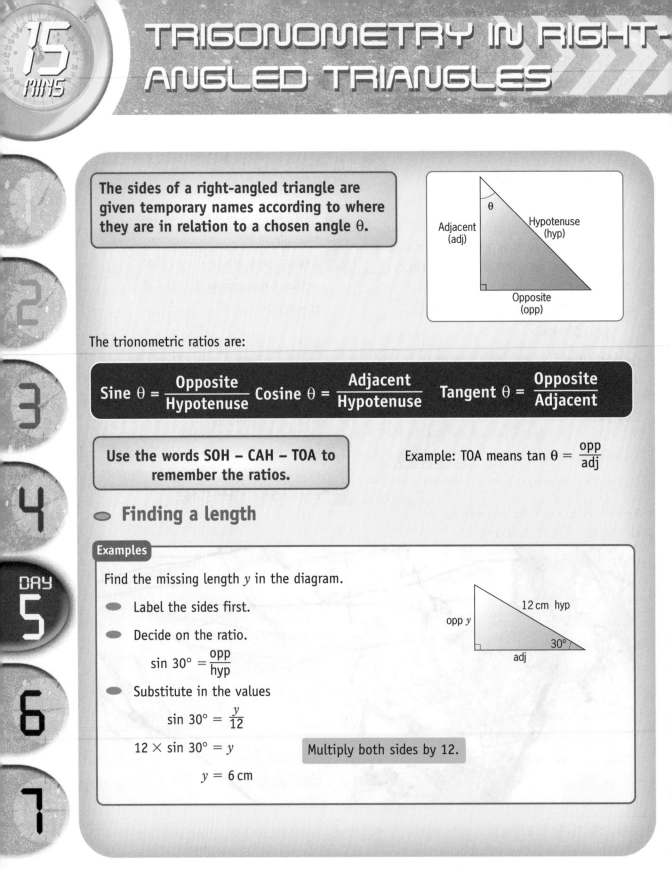

The trionometric ratios are:

$$\text{Sine } \theta = \frac{\text{Opposite}}{\text{Hypotenuse}} \quad \text{Cosine } \theta = \frac{\text{Adjacent}}{\text{Hypotenuse}} \quad \text{Tangent } \theta = \frac{\text{Opposite}}{\text{Adjacent}}$$

Use the words SOH – CAH – TOA to remember the ratios.

Example: TOA means $\tan \theta = \frac{\text{opp}}{\text{adj}}$

Finding a length

Examples

Find the missing length y in the diagram.

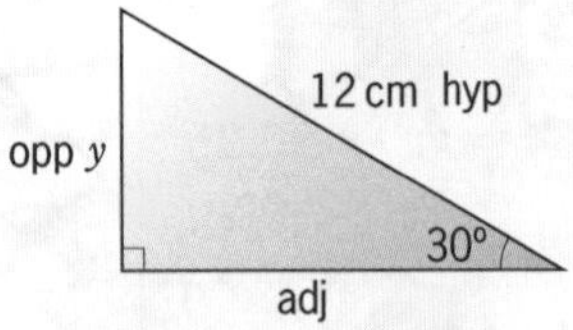

- Label the sides first.
- Decide on the ratio.

$\sin 30° = \frac{\text{opp}}{\text{hyp}}$

- Substitute in the values

$\sin 30° = \frac{y}{12}$

$12 \times \sin 30° = y$ — Multiply both sides by 12.

$y = 6\text{ cm}$

DAY 5

Trigonometry in right-angled triangles can be used to find an unknown angle or length.

Finding an angle

Examples

Calculate angle $A\hat{B}C$.

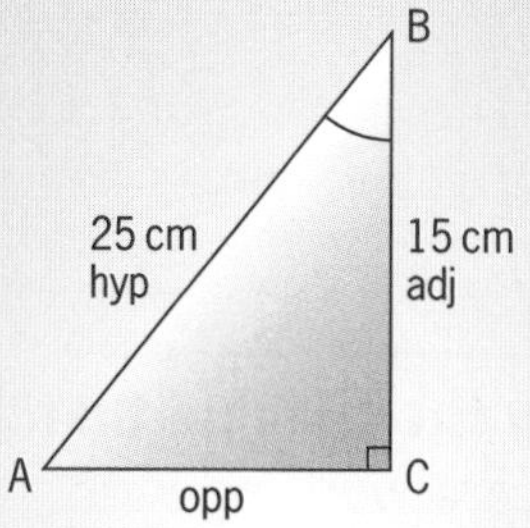

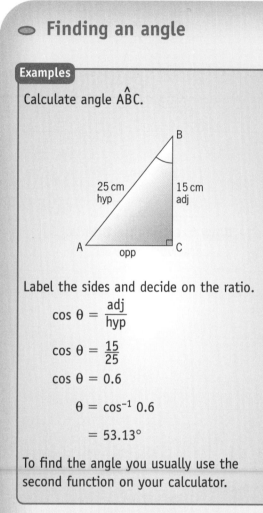

Label the sides and decide on the ratio.

$$\cos\theta = \frac{\text{adj}}{\text{hyp}}$$

$$\cos\theta = \frac{15}{25}$$

$$\cos\theta = 0.6$$

$$\theta = \cos^{-1} 0.6$$

$$= 53.13°$$

To find the angle you usually use the second function on your calculator.

It is important you know how to use your calculator when working out trigonometry questions.

PLEASURE PARK
Drop Zone here I come!
The angle of elevation is 75°.
The height of the vertical drop is 40 metres, so what is the ....
Aaaaagh !

TRIGONOMETRY IN 3-D DIAGRAMS

Worked example

A, C and D are three points on horizontal ground.
A is 1600 m due west of D and C is 1420 m due south of D.

BD is a vertical tower.

The angle of elevation of B from C is 21°.

a) Calculate the height of the tower.

b) Find the angle of elevation of B from A.

V is a point on AC which is nearest to D.

c) Calculate the angle of elevation of B from V.

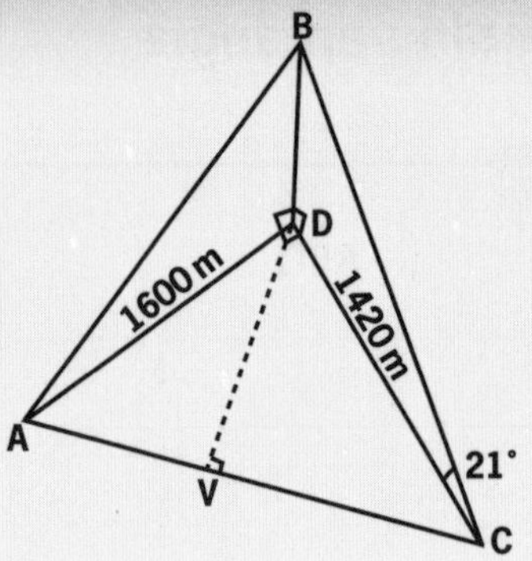

Draw the triangle that you are working with.

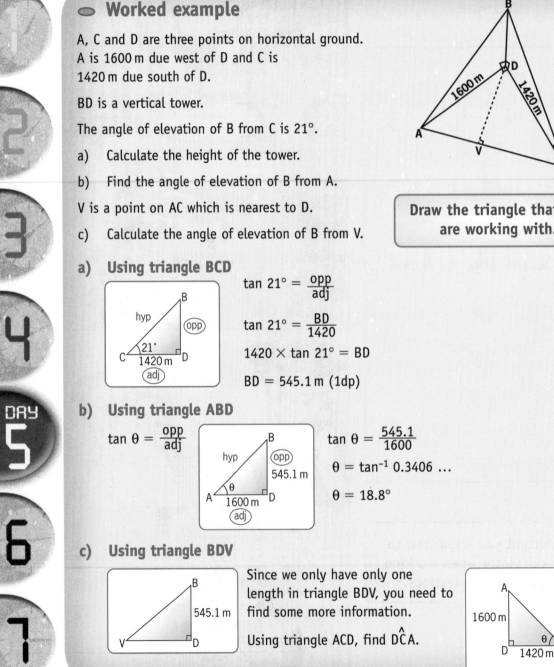

a) Using triangle BCD

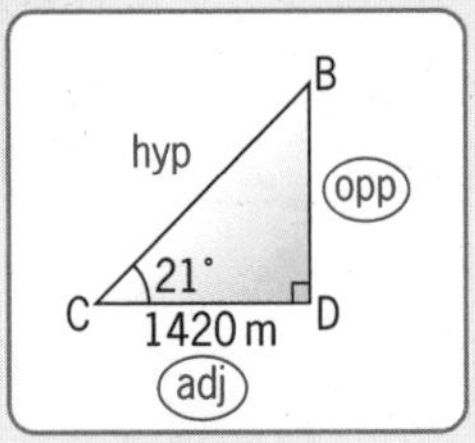

$\tan 21° = \frac{\text{opp}}{\text{adj}}$

$\tan 21° = \frac{BD}{1420}$

$1420 \times \tan 21° = BD$

$BD = 545.1\text{ m (1dp)}$

b) Using triangle ABD

$\tan \theta = \frac{\text{opp}}{\text{adj}}$

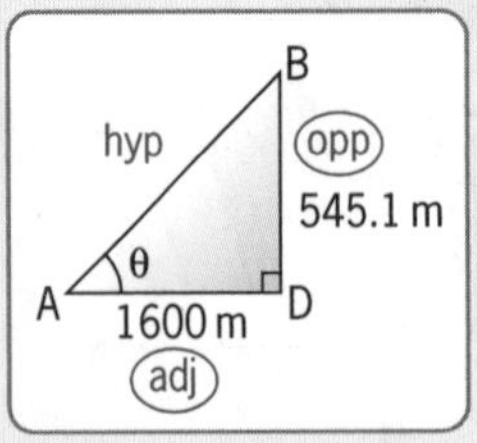

$\tan \theta = \frac{545.1}{1600}$

$\theta = \tan^{-1} 0.3406 \ldots$

$\theta = 18.8°$

c) Using triangle BDV

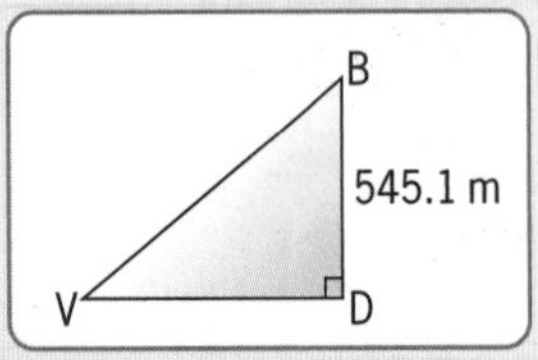

Since we only have only one length in triangle BDV, you need to find some more information.

Using triangle ACD, find $D\hat{C}A$.

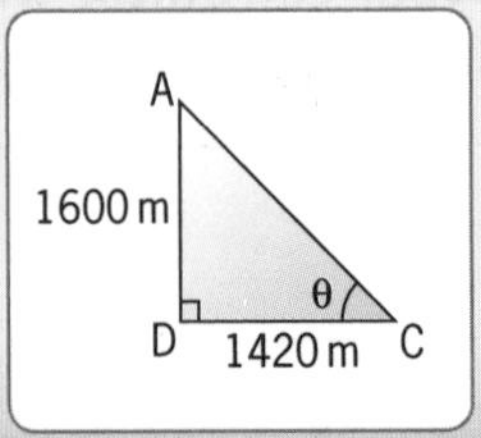

The use of trigonometry and Pythagoras' theorem can be applied when solving problems in three-dimensional figures.

$\tan\theta = \frac{1600}{1420}$

$\theta = \tan^{-1} 1.126\ldots$

$\theta = 48.41°$

Now use triangle DVC to find length DV.

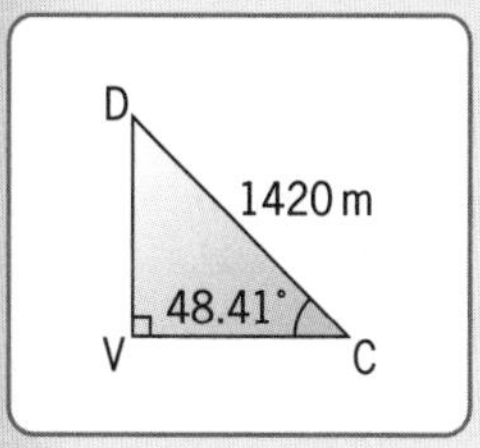

$\sin 48.41° = \frac{DV}{1420}$

$DV = \sin 48.41° \times 1420$

$DV = 1062.05\,\text{m}$

Now we use triangle BDV to find angle BVD.

$\tan\theta = \frac{545.1}{1062.05}$

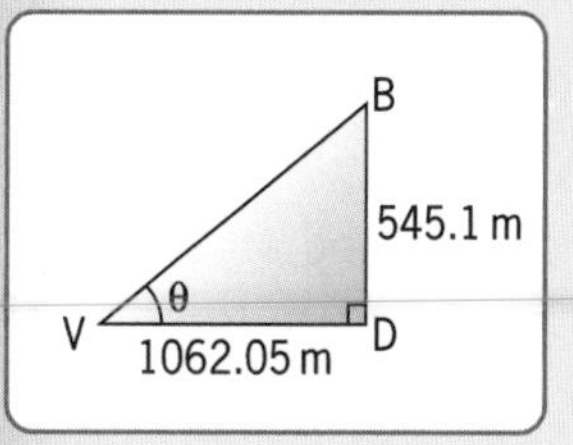

$\theta = \tan^{-1} 0.513\ldots$

$\theta = 27.2°$ (2dp)

This is an example of a multistepped question in that you have to do several parts to get to the answer. They are usually worth lots of marks.

Progress check

The diagram shows a cuboid.

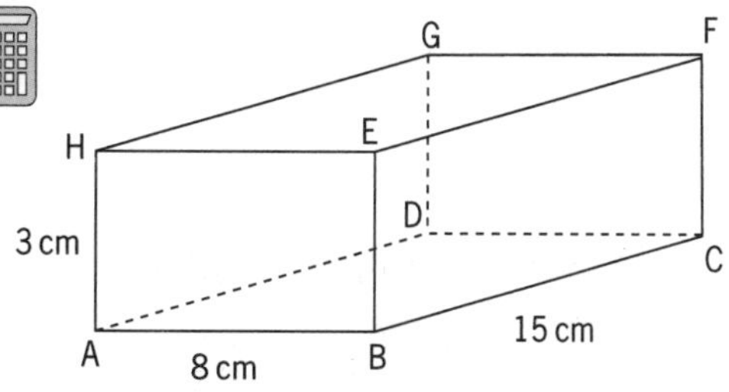

1 Decide whether the following statements are true or false.

a) The length HB is 8.54 cm (2dp)

b) The length HC is 18 cm

c) Angle ACD is 47°

d) Angle HBA is 20.6° (1dp)

2 Calculate the size of angle HCA.

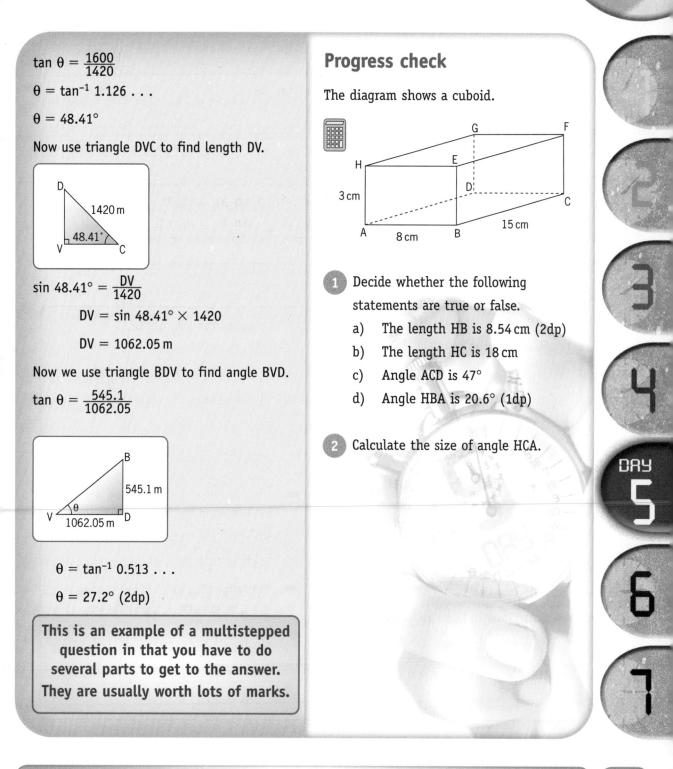

THE SINE AND COSINE RULES

The sine rule

The standard ways to write down the sine and cosine rule use the following notation for sides and angles in a general triangle.

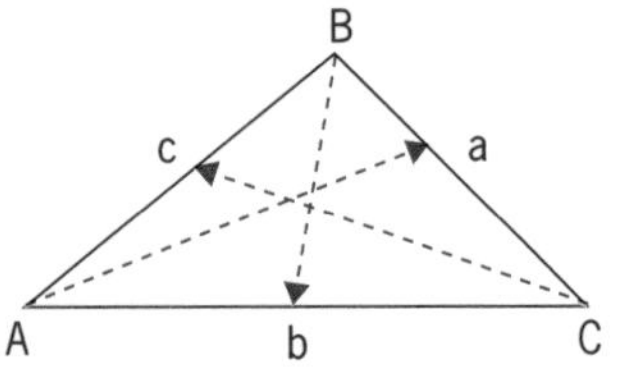

When finding a length:

$$\frac{a}{\sin A} = \frac{b}{\sin B} = \frac{c}{\sin C}$$

When finding an angle:

$$\frac{\sin A}{a} = \frac{\sin B}{b} = \frac{\sin C}{c}$$

Examples

1. Calculate the length of RS.

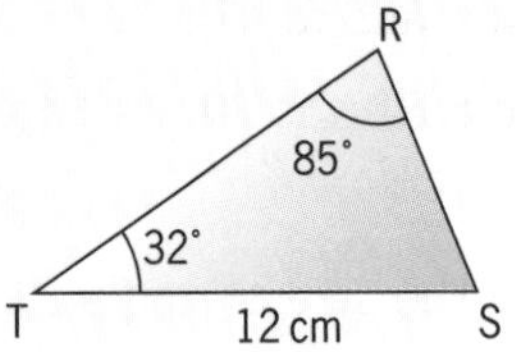

- Call RS, x, the length to be found.
- Since we have two angles and a side we use the sine rule

$$\frac{x}{\sin 32°} = \frac{12}{\sin 85°}$$

- Rearrange to make x the subject:

$$x = \frac{12}{\sin 85°} \times \sin 32°$$

Length of RS = 6.38 cm (3sf)

2. Calculate the size of $C\hat{D}E$.

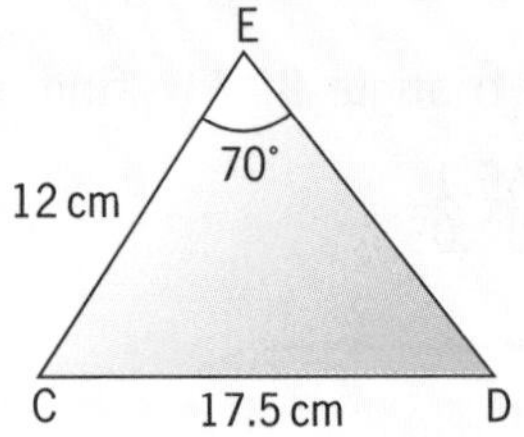

- Since we have two sides and an angle not enclosed by them. we use the sine rule.

$$\frac{\sin D}{12} = \frac{\sin 70°}{17.5}$$

- Rearrange to give:

$$\sin D = \frac{\sin 70°}{17.5} \times 12$$

$$\sin D = 0.644\ldots$$

$$D = \sin^{-1} 0.644$$

$$D = 40.1°$$

The sine and cosine rules allow you to solve problems in triangles that do not contain a right angle.

20 MINS

The cosine rule

When finding a length:
$$a^2 = b^2 + c^2 - (2bc \cos A)$$
When finding an angle:
$$\cos A = \frac{b^2 + c^2 - a^2}{2bc}$$

Example

Calculate the length of JK.

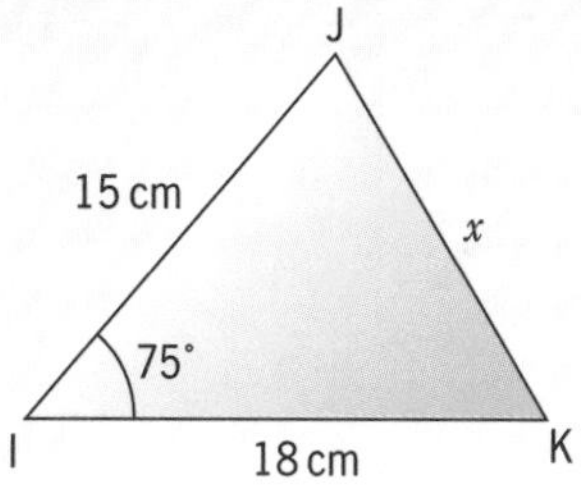

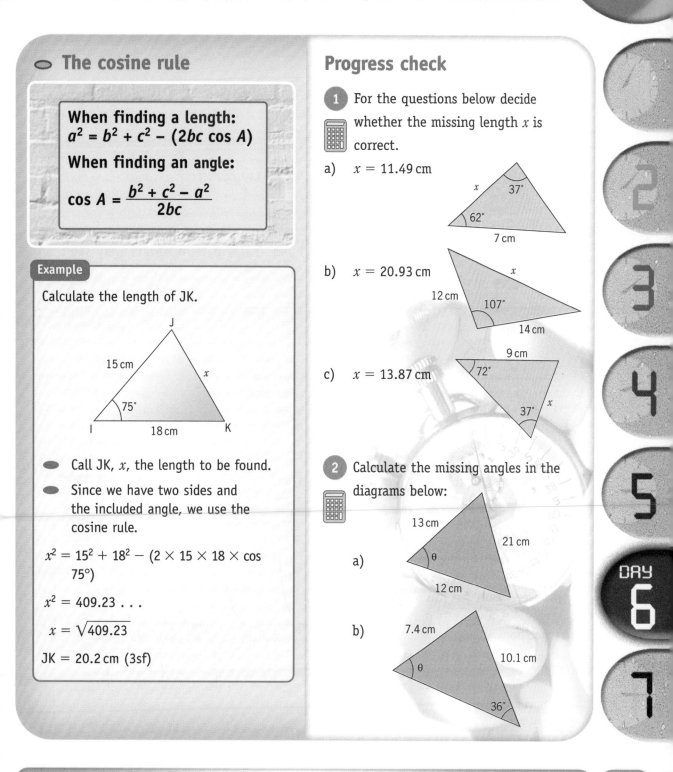

- Call JK, x, the length to be found.
- Since we have two sides and the included angle, we use the cosine rule.

$x^2 = 15^2 + 18^2 - (2 \times 15 \times 18 \times \cos 75°)$

$x^2 = 409.23\ldots$

$x = \sqrt{409.23}$

JK = 20.2 cm (3sf)

Progress check

1. For the questions below decide whether the missing length x is correct.

a) $x = 11.49$ cm

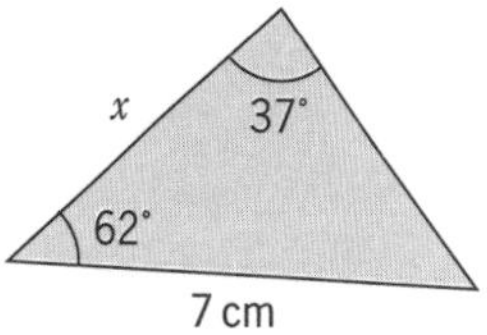

b) $x = 20.93$ cm

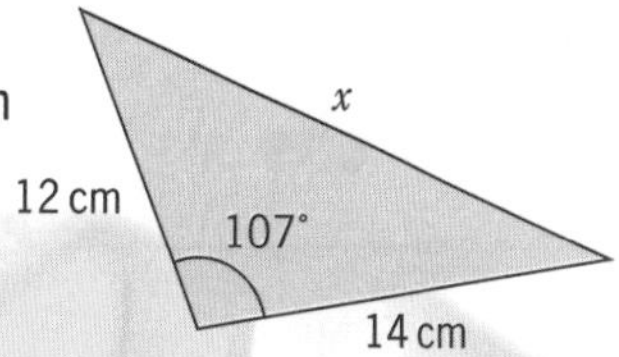

c) $x = 13.87$ cm

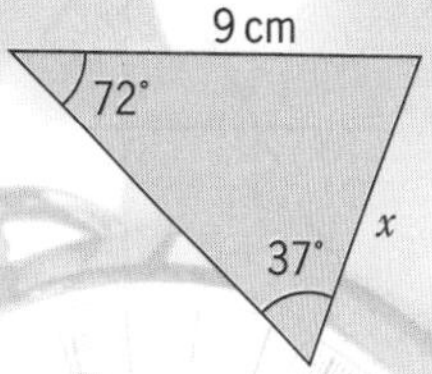

2. Calculate the missing angles in the diagrams below:

a)

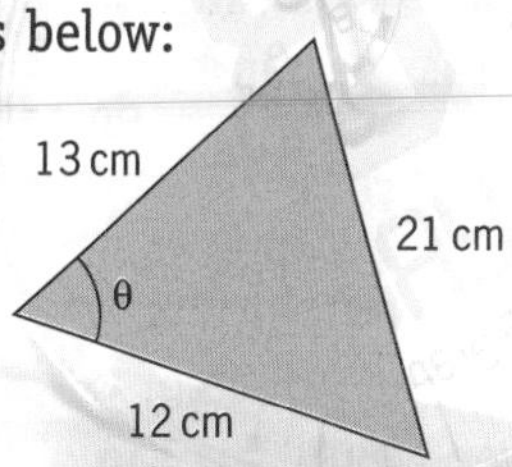

b)

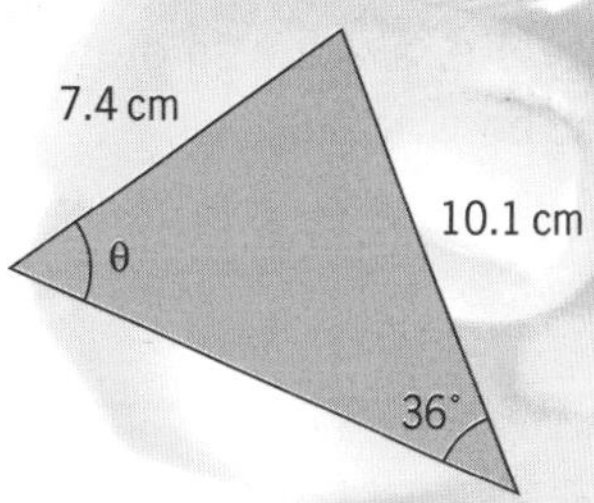

DAY 6

TRIGONOMETRIC FUNCTIONS

$y = \sin x$

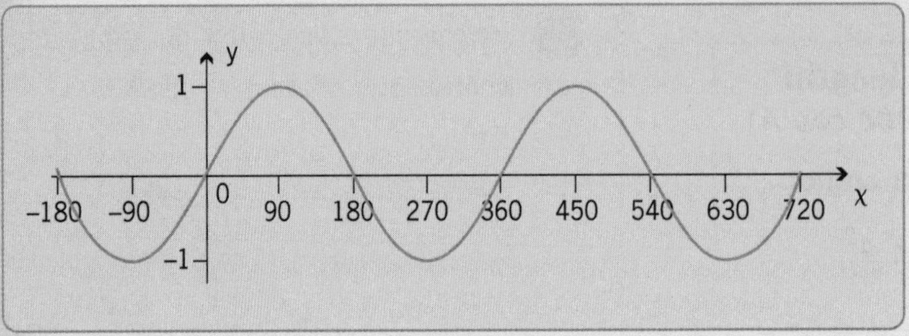

The maximum and minimum values of sin x are 1 and –1. The pattern repeats every 360°.

$y = \cos x$

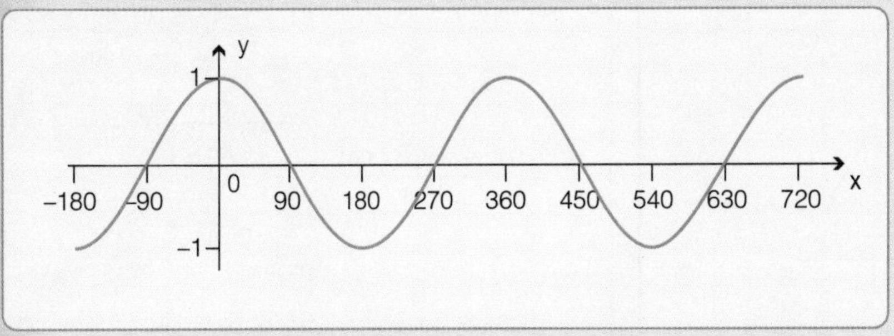

The maximum and minimum values of cos x are 1 and –1. The pattern repeats every 360°. This graph is the same as $y = \sin x$ except it has been moved 90° to the left.

$y = \tan x$

This graph is nothing like the two above. The values of tan x repeat every 180°. The tan of 90° is infinity, i.e. a value so great it cannot be written down.

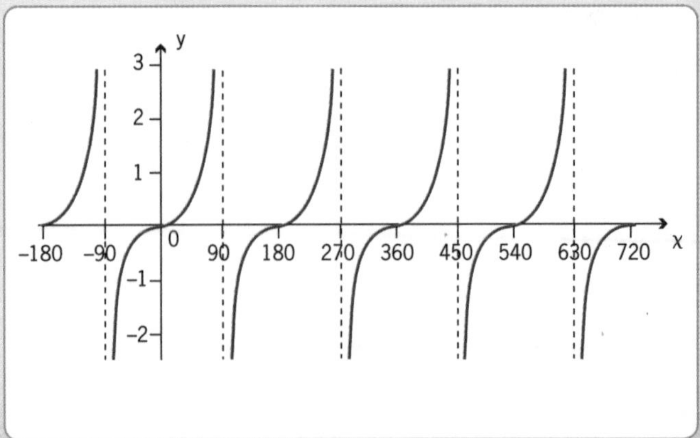

The behaviour of the sine, cosine and tangent functions may be represented graphically.

10 MINS

The trigonometric graphs can be used to solve inverse problems.

Example

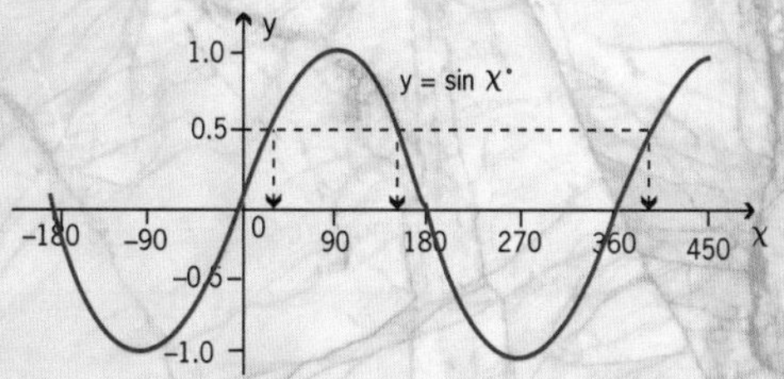

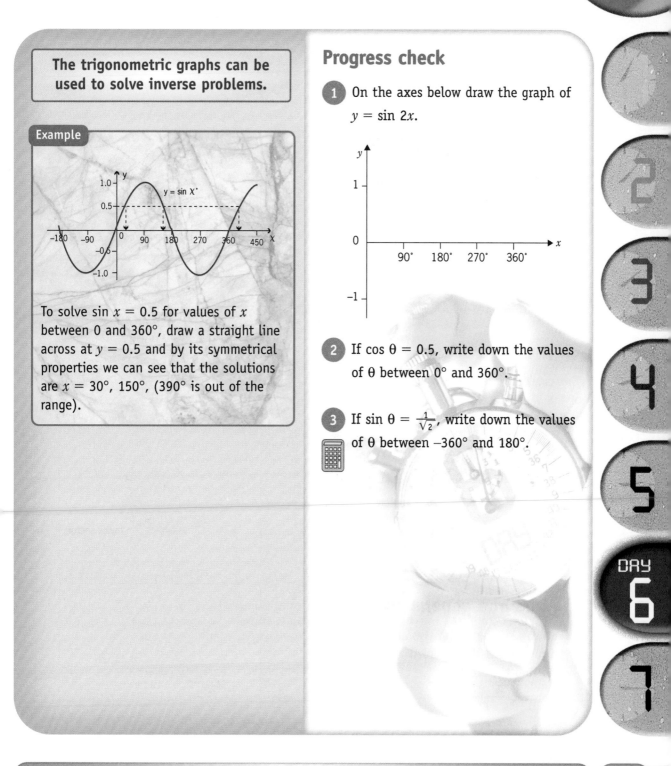

To solve sin $x = 0.5$ for values of x between 0 and 360°, draw a straight line across at $y = 0.5$ and by its symmetrical properties we can see that the solutions are $x = 30°$, 150°, (390° is out of the range).

Progress check

1. On the axes below draw the graph of $y = \sin 2x$.

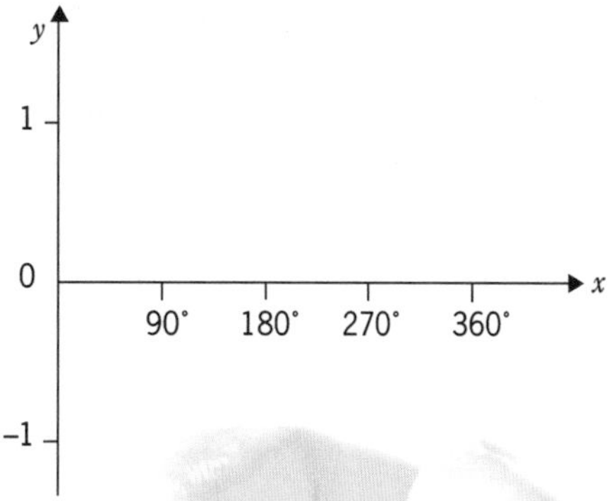

2. If $\cos\theta = 0.5$, write down the values of θ between 0° and 360°.

3. If $\sin\theta = \frac{1}{\sqrt{2}}$, write down the values of θ between –360° and 180°.

ARC, SECTOR AND SEGMENT

15 MINS

Length of a circular arc

This can be expressed as a fraction of the circumference of a circle.

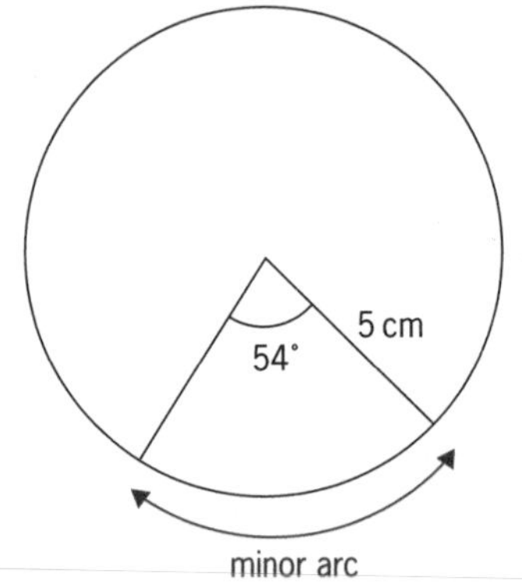

$$\text{Arc length} = \frac{\theta}{360^\circ} \times 2\pi r$$

where θ is the angle subtended at the centre.

$= \frac{54^\circ}{360^\circ} \times 2 \times \pi \times 5$

$= 4.71\text{ cm (2dp)}$

Area of a sector

This can be expressed as the fraction of the area of a circle.

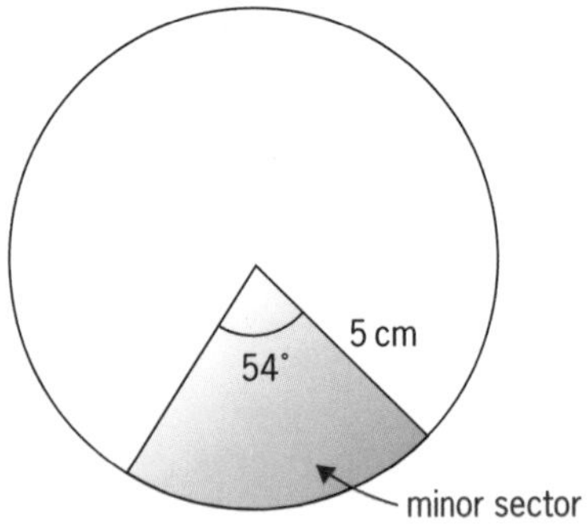

$$\text{Area of sector} = \frac{\theta}{360^\circ} \times \pi r^2$$

$= \frac{54^\circ}{360^\circ} \times \pi \times 5^2$

$= 11.78\text{ cm}^2\text{ (2dp)}$

Area of a general triangle

If we know the length of two sides of a triangle and the included angle, we can find the area.

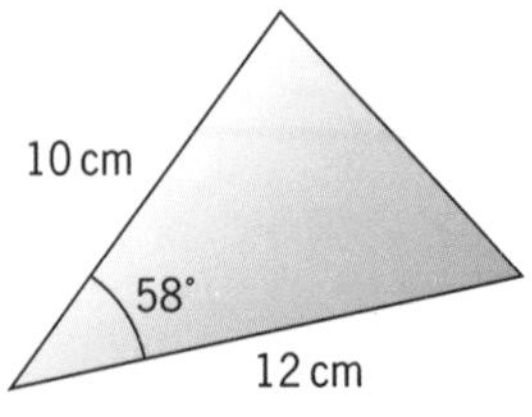

DAY 6

The length of an arc is a fraction of the circumference. The area of a sector is a fraction of the area of a circle.

Area = $\frac{1}{2} \times b \times c \times \sin A$

two lengths ($b \times c$) — included angle (A)

This formula is given on the exam paper.

Area $= \frac{1}{2} \times 10 \times 12 \times \sin 58°$

$= 50.88\,\text{cm}^2$ (2dp)

Area of a segment

This can be worked out in two stages:

1. Calculate the area of the sector and the area of the triangle.
2. Subtract the area of the triangle from the area of the sector.

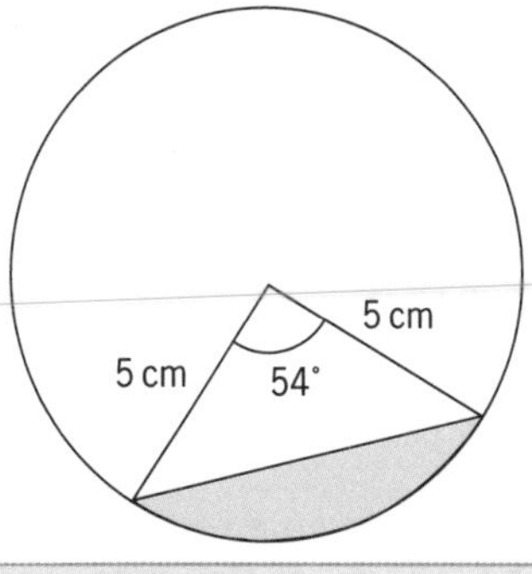

Segment area =

$\left(\frac{54°}{360°} \times \pi \times 5^2\right) - \left(\frac{1}{2} \times 5 \times 5 \times \sin 54°\right)$

$= 11.78 \ldots - 10.11 \ldots$

$= 1.67\,\text{cm}^2$ (2dp)

Progress check

1. For this diagram, calculate:
   a) the arc length (x)
   b) the sector area
   c) the area of the shaded segment

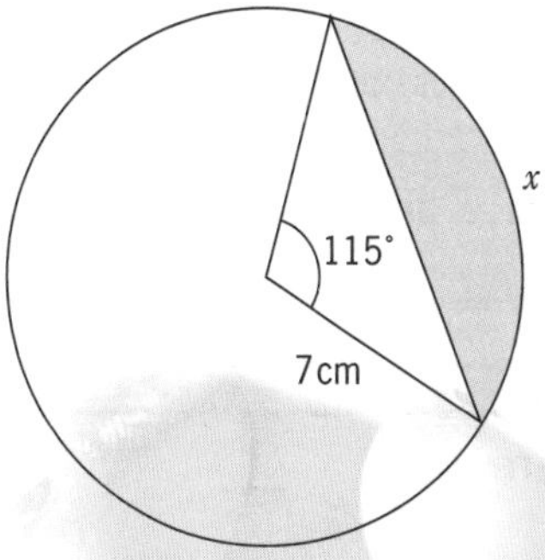

2. Decide whether this statement is true or false. You must show sufficient working out to justify your answer.
   The area of the shaded segment is 9.03 cm².

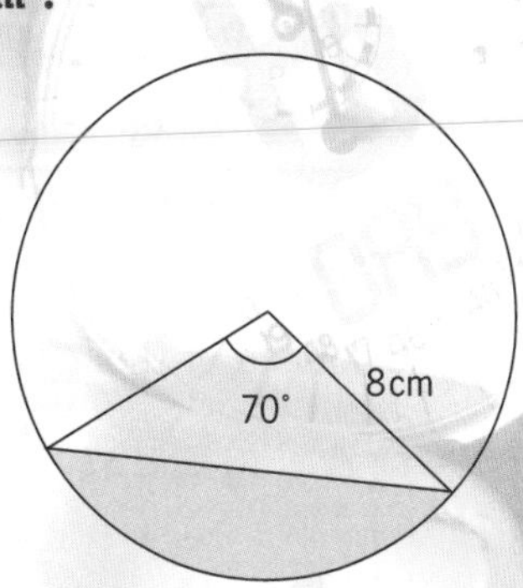

SURFACE AREA AND VOLUME

Key formulae

Volume of a prism

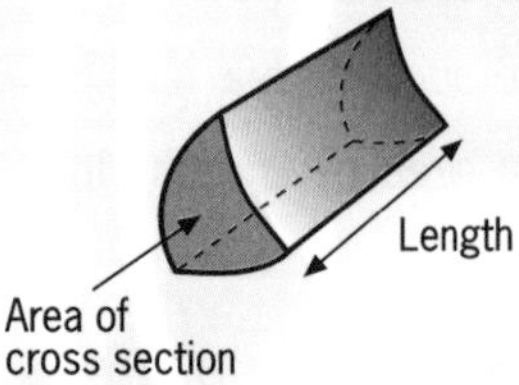

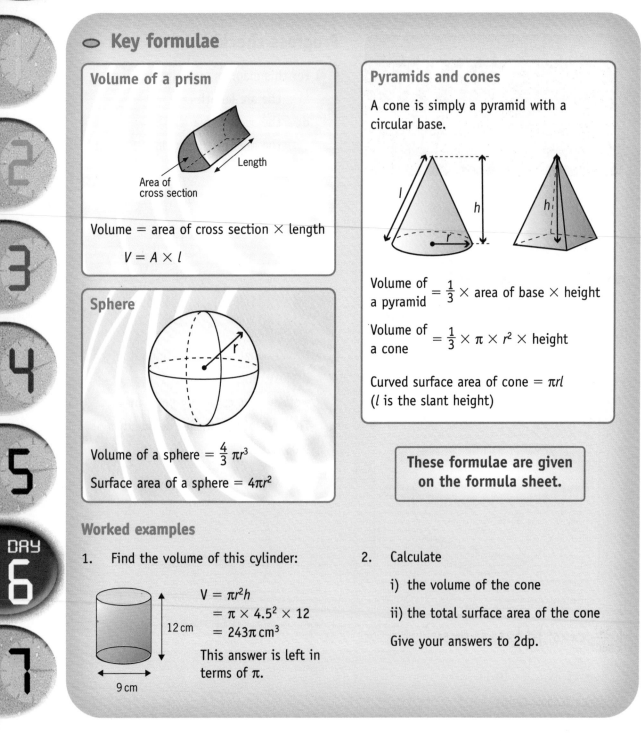

Volume = area of cross section × length

$V = A \times l$

Sphere

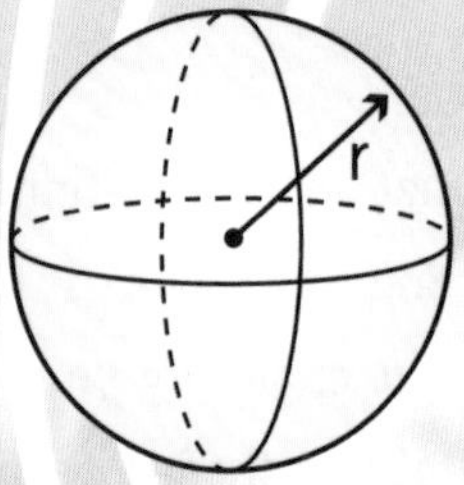

Volume of a sphere $= \frac{4}{3}\pi r^3$

Surface area of a sphere $= 4\pi r^2$

Pyramids and cones

A cone is simply a pyramid with a circular base.

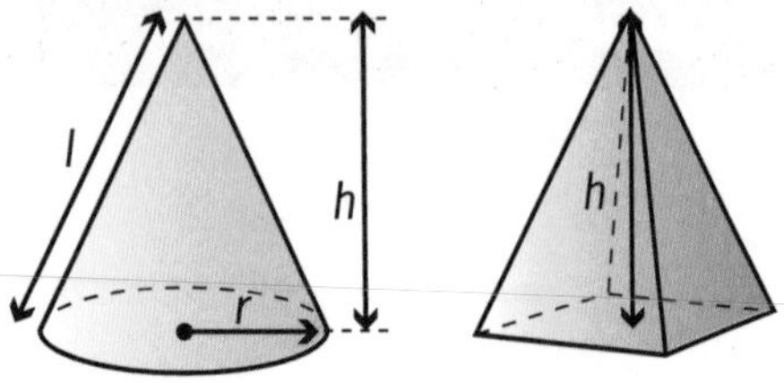

Volume of a pyramid $= \frac{1}{3} \times$ area of base × height

Volume of a cone $= \frac{1}{3} \times \pi \times r^2 \times$ height

Curved surface area of cone $= \pi r l$
(l is the slant height)

These formulae are given on the formula sheet.

Worked examples

1. Find the volume of this cylinder:

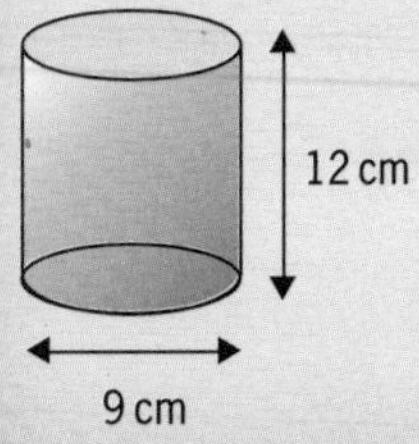

$V = \pi r^2 h$
$= \pi \times 4.5^2 \times 12$
$= 243\pi \text{ cm}^3$

This answer is left in terms of π.

2. Calculate
   i) the volume of the cone
   ii) the total surface area of the cone

   Give your answers to 2dp.

A prism is any solid that can be cut up into slices that are all the same shape. This is known as having a uniform cross-section.

15 MINS

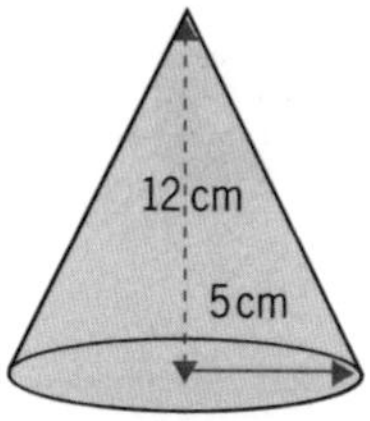

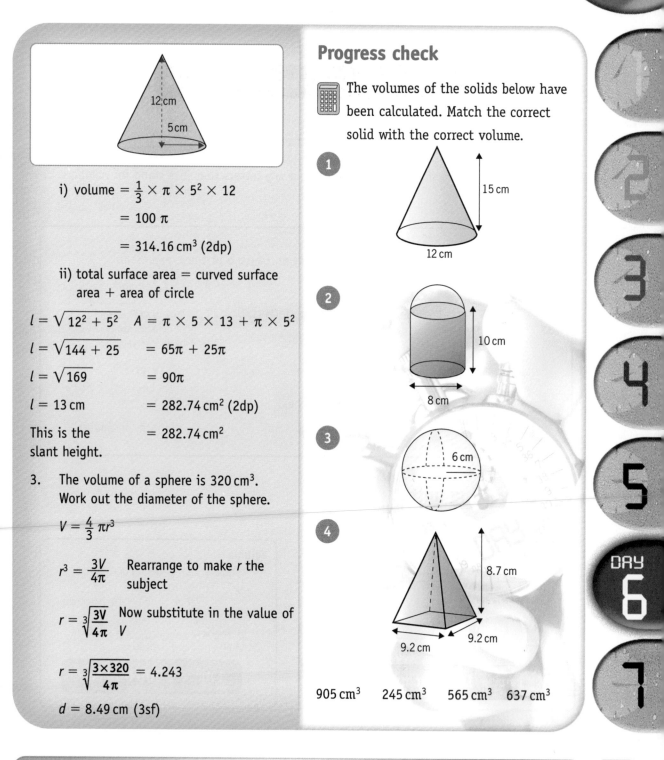

i) volume $= \frac{1}{3} \times \pi \times 5^2 \times 12$

$= 100\pi$

$= 314.16\ \text{cm}^3$ (2dp)

ii) total surface area = curved surface area + area of circle

$l = \sqrt{12^2 + 5^2}$	$A = \pi \times 5 \times 13 + \pi \times 5^2$
$l = \sqrt{144 + 25}$	$= 65\pi + 25\pi$
$l = \sqrt{169}$	$= 90\pi$
$l = 13$ cm	$= 282.74\ \text{cm}^2$ (2dp)
This is the slant height.	$= 282.74\ \text{cm}^2$

3. The volume of a sphere is $320\ \text{cm}^3$. Work out the diameter of the sphere.

$V = \frac{4}{3}\pi r^3$

$r^3 = \frac{3V}{4\pi}$ Rearrange to make r the subject

$r = \sqrt[3]{\frac{3V}{4\pi}}$ Now substitute in the value of V

$r = \sqrt[3]{\frac{3 \times 320}{4\pi}} = 4.243$

$d = 8.49$ cm (3sf)

Progress check

The volumes of the solids below have been calculated. Match the correct solid with the correct volume.

1

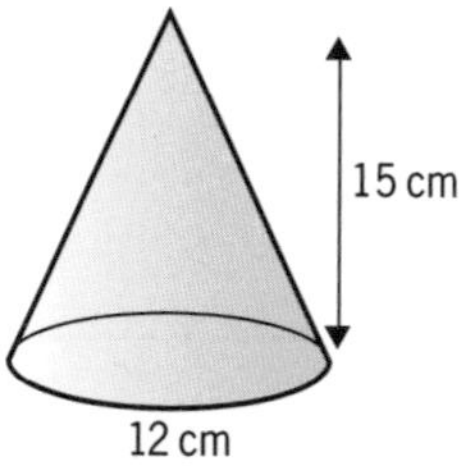

2

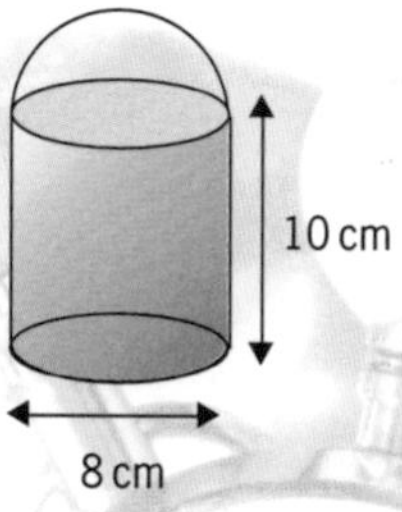

3

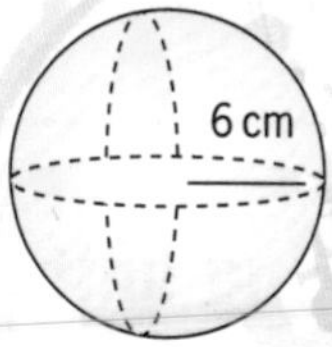

4

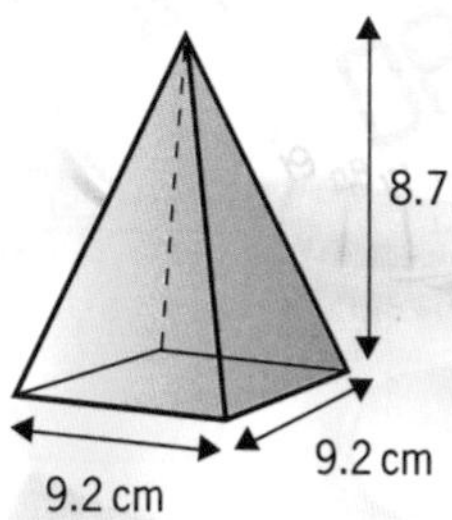

$905\ \text{cm}^3$ $245\ \text{cm}^3$ $565\ \text{cm}^3$ $637\ \text{cm}^3$

DIMENSIONS AND CONVERTING UNITS

Dimensions

L = length L^2 = area L^3 = volume

- A formula with a mixed dimension is impossible, e.g. $L^2 + L^3$.
- A dimension greater than 3 is impossible.
- Null quantities are numbers, including π and any letters that just stand for numbers, fractions etc.

Examples

If a, b, c represent lengths, what is the dimension of each of these expressions?

1. $\frac{2}{3}a^2 + 4\pi b^2$

$\rightarrow \frac{2}{3}L^2 + 4\pi L^2$	Change to dimension letters.
$\rightarrow L^2 + L^2$	Remove null quantities.
$\rightarrow 2L^2$	Simplify.
$\rightarrow L^2$	Remove null quantities.
(area)	Decide on dimension.

2. $\sqrt{4c^2 + 6a^2}$

$\rightarrow \sqrt{4L^2 + 6L^2}$

$\rightarrow \sqrt{L^2 + L^2}$

$\rightarrow \sqrt{2L^2}$

$\rightarrow \sqrt{L^2} \rightarrow L$

(length/perimeter)

Converting units

You need to remember these facts:

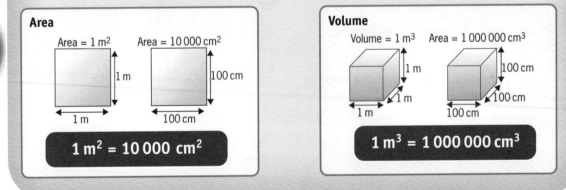

Area

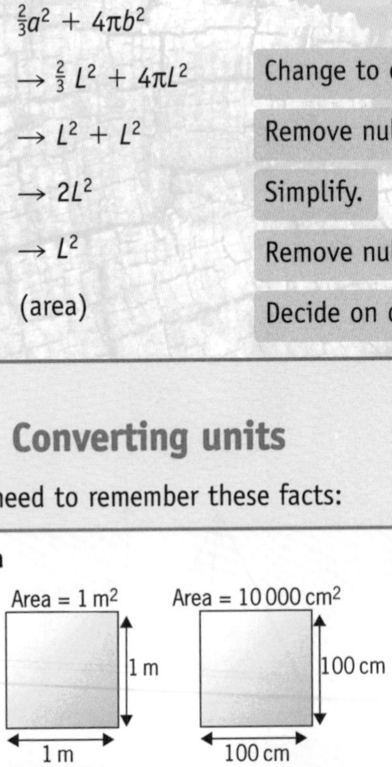

$1\ m^2 = 10\,000\ cm^2$

Volume

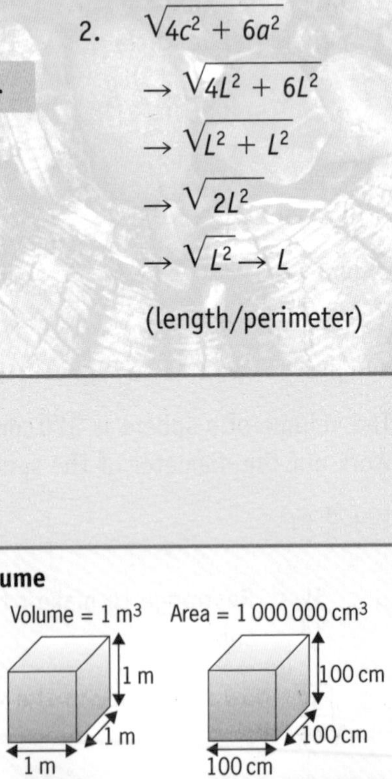

$1\ m^3 = 1\,000\,000\ cm^3$

All types of measurement are built up from basic dimensions of length, mass and time.

Remember

When a question has different units, e.g.

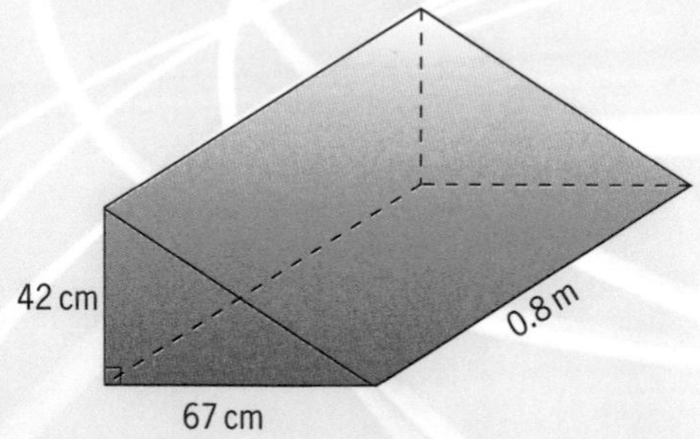

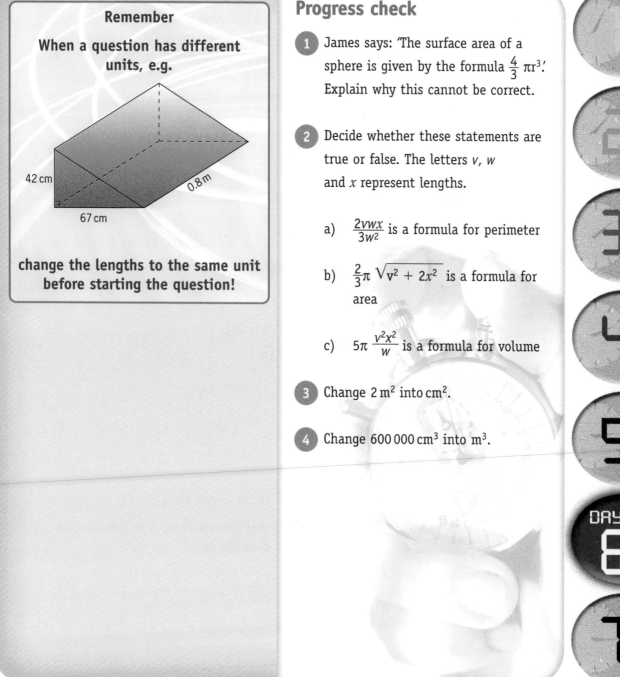

change the lengths to the same unit before starting the question!

Progress check

1. James says: 'The surface area of a sphere is given by the formula $\frac{4}{3}\pi r^3$.' Explain why this cannot be correct.

2. Decide whether these statements are true or false. The letters v, w and x represent lengths.

   a) $\frac{2vwx}{3w^2}$ is a formula for perimeter

   b) $\frac{2}{3}\pi\sqrt{v^2 + 2x^2}$ is a formula for area

   c) $5\pi\frac{v^2x^2}{w}$ is a formula for volume

3. Change 2 m^2 into cm^2.

4. Change 600 000 cm^3 into m^3.

VECTORS 1

Four types of notation are used to represent vectors. The vector shown here can be referred to as

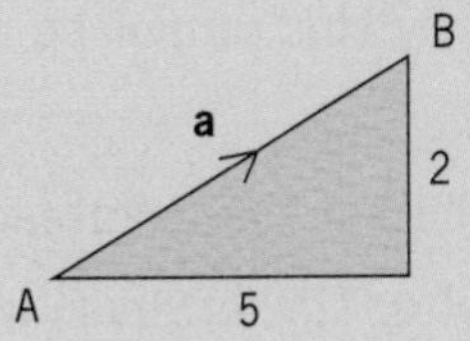

$\binom{5}{2}$ or $\underline{a}$ or $\overrightarrow{AB}$ or $\mathbf{a}$

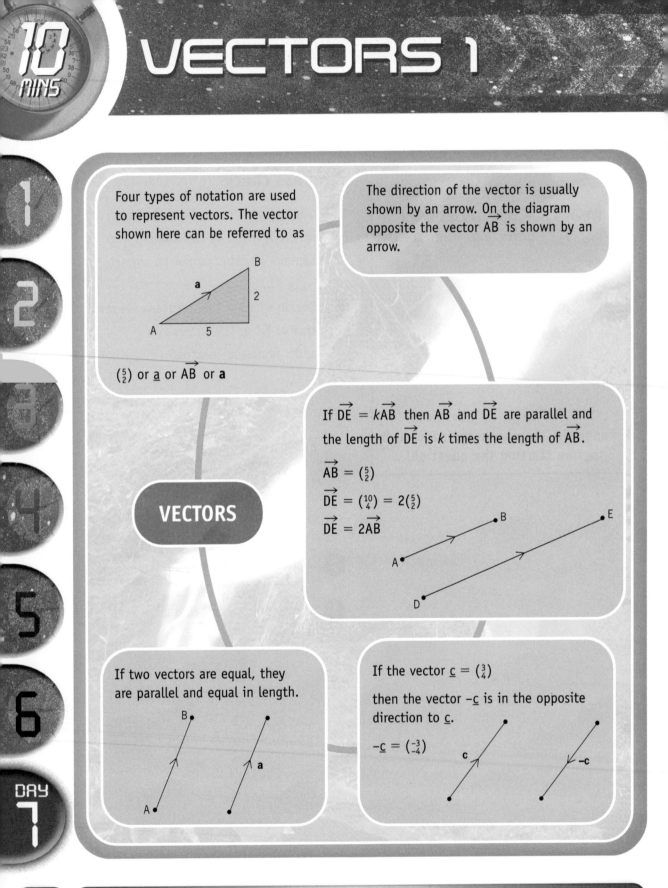

The direction of the vector is usually shown by an arrow. On the diagram opposite the vector $\overrightarrow{AB}$ is shown by an arrow.

VECTORS

If $\overrightarrow{DE} = k\overrightarrow{AB}$ then $\overrightarrow{AB}$ and $\overrightarrow{DE}$ are parallel and the length of $\overrightarrow{DE}$ is k times the length of $\overrightarrow{AB}$.

$\overrightarrow{AB} = \binom{5}{2}$

$\overrightarrow{DE} = \binom{10}{4} = 2\binom{5}{2}$

$\overrightarrow{DE} = 2\overrightarrow{AB}$

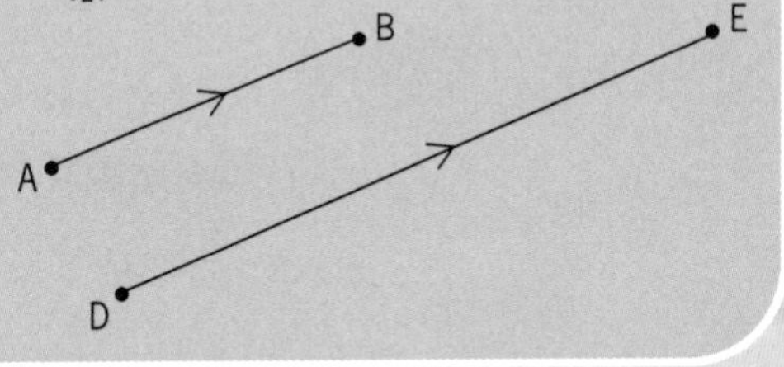

If two vectors are equal, they are parallel and equal in length.

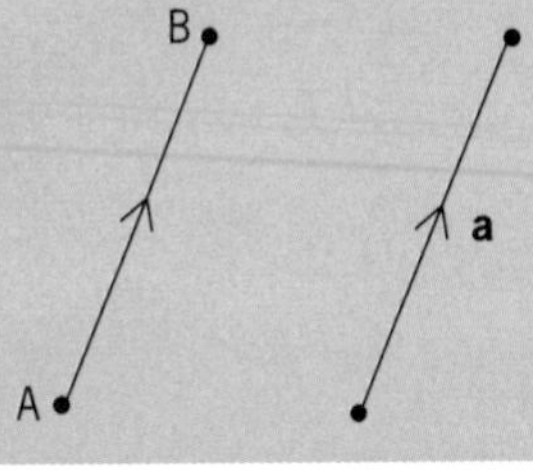

If the vector $\underline{c} = \binom{3}{4}$

then the vector $-\underline{c}$ is in the opposite direction to $\underline{c}$.

$-\underline{c} = \binom{-3}{-4}$

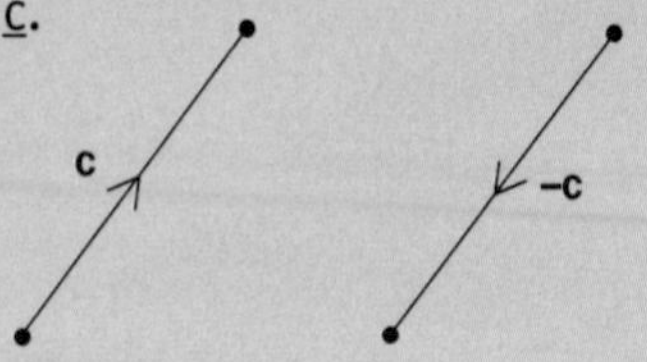

A vector is a quantity that has both distance and direction.

Magnitude of a vector

The magnitude of a vector is the length of the directed line segment representing it.

Pythagoras' theorem can be used to find the magnitude.

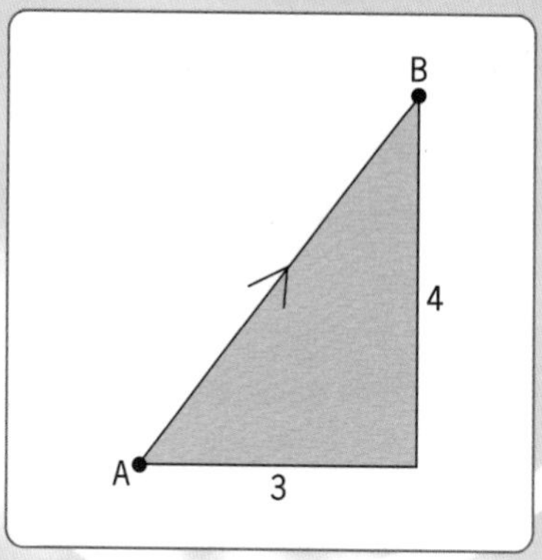

In general, the magnitude of a vector $\binom{x}{y}$ is $\sqrt{x^2 + y^2}$

The magnitude of AB

$(AB)^2 = 3^2 + 4^2$

$(AB)^2 = 9 + 16$

$(AB)^2 = 25$

$AB = \sqrt{25}$

$AB = 5$

The magnitude of AB is 5 units.

Splitting vectors into components

- Any vector can be split up into two components that are at 90° to each other.
- These two components will be ***F*cos**θ and ***F*sin**θ

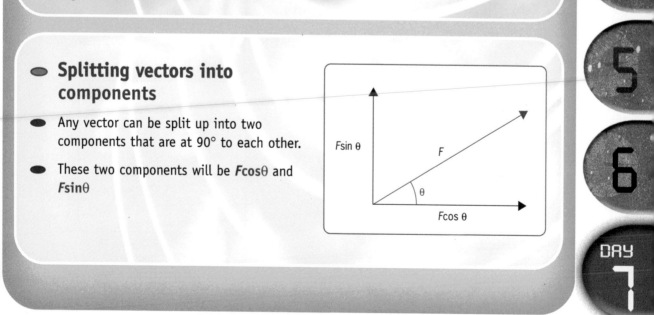

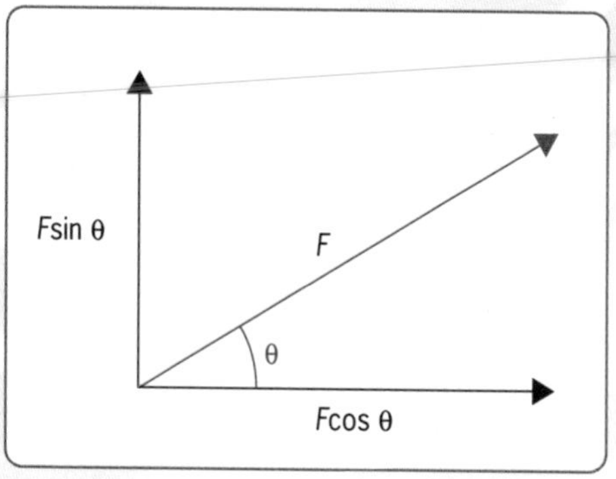

VECTORS 2

Addition and subtraction of vectors

- The **resultant** of two vectors is found by adding them.
- Vectors must always be added end to end so that the arrows follow on from each other.
- A resultant is usually labelled with a double arrow.

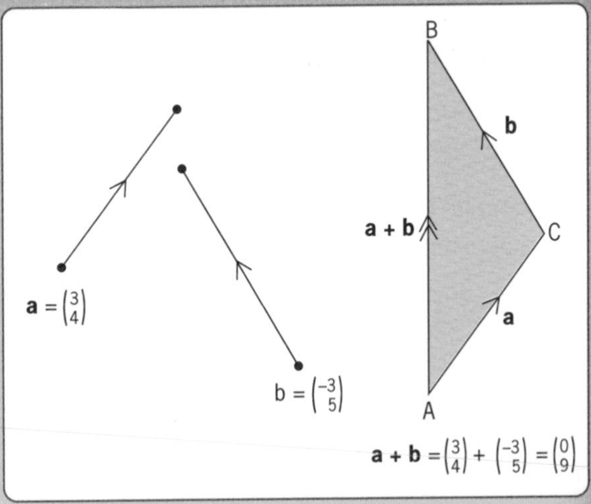

This shows the triangle law of addition. To take the route directly from A to B, is equivalent to travelling via C, hence we can represent $\overrightarrow{AB}$ as **a** + **b**.

Vectors can also be subtracted.

a − **b** can be interpreted as **a** + (−**b**).

$\mathbf{a} + (-\mathbf{b}) = \begin{pmatrix} 3 \\ 4 \end{pmatrix} + \begin{pmatrix} 3 \\ -5 \end{pmatrix} \therefore \mathbf{a} - \mathbf{b} = \begin{pmatrix} 6 \\ -1 \end{pmatrix}$

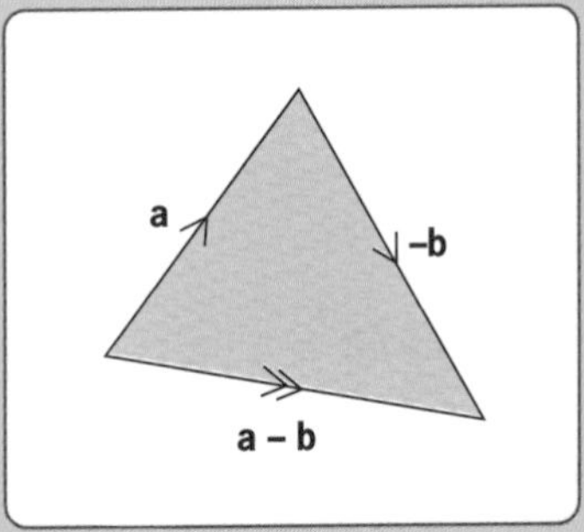

Position vectors

- The position vector of a point Q is the vector $\overrightarrow{OQ}$, where O is the origin.

In the diagram the position vectors of B and C are **b and c** respectively. Using this notation:

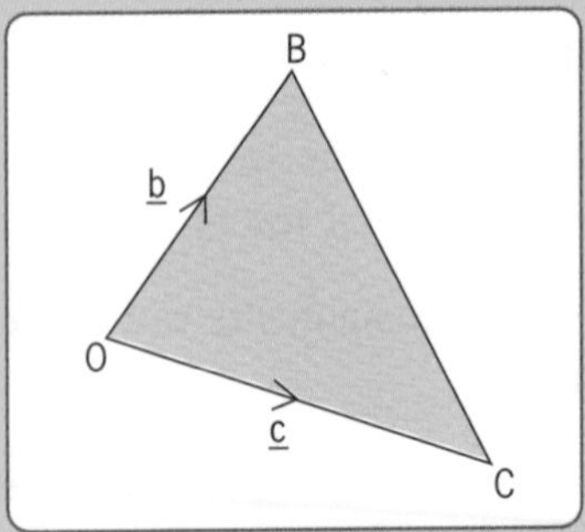

$\overrightarrow{BC} = -\underline{b} + \underline{c}$ or

$\underline{c} - \underline{b}$

Example

OAB is a triangle. Given that $\overrightarrow{OA} = \underline{a}$, $\overrightarrow{OB} = \underline{b}$ and that N splits $\overrightarrow{AB}$ in the ratio 1 : 2, find in terms of $\underline{a}$ and $\underline{b}$ the vectors:

i) $\overrightarrow{AB}$ ii) $\overrightarrow{ON}$

i) $\overrightarrow{AB} = \overrightarrow{OA} + \overrightarrow{OB}$
(go from A to B via O)

$= -\underline{a} + \underline{b}$

ii) $\overrightarrow{ON} = \overrightarrow{OA} + \overrightarrow{AN} \quad (\overrightarrow{AN} = \frac{1}{3}\overrightarrow{AB})$

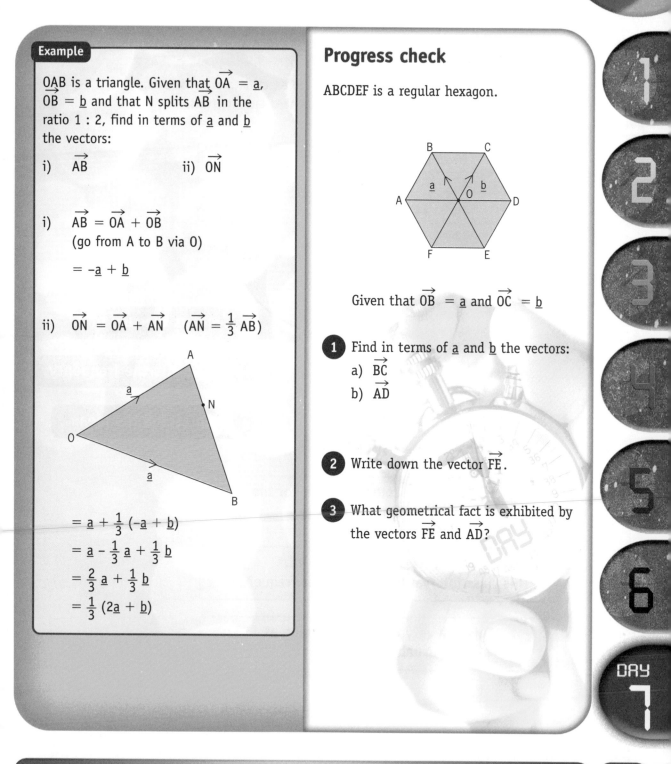

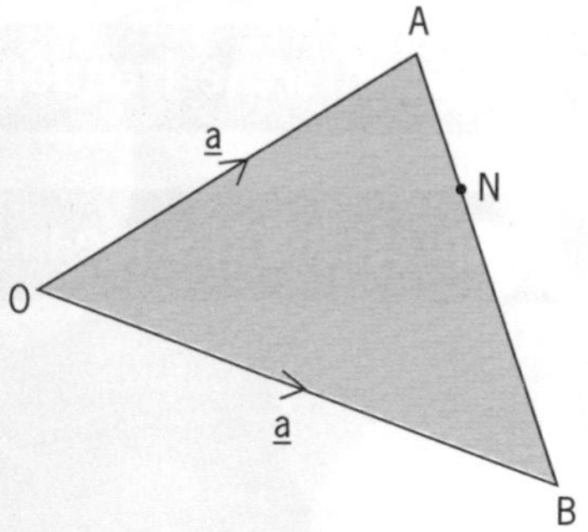

$= \underline{a} + \frac{1}{3}(-\underline{a} + \underline{b})$

$= \underline{a} - \frac{1}{3}\underline{a} + \frac{1}{3}\underline{b}$

$= \frac{2}{3}\underline{a} + \frac{1}{3}\underline{b}$

$= \frac{1}{3}(2\underline{a} + \underline{b})$

Progress check

ABCDEF is a regular hexagon.

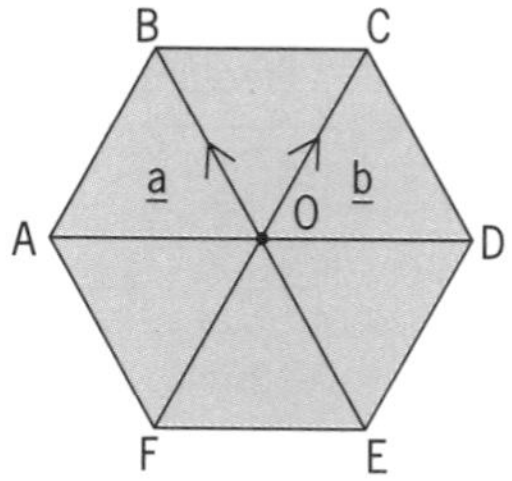

Given that $\overrightarrow{OB} = \underline{a}$ and $\overrightarrow{OC} = \underline{b}$

1. Find in terms of $\underline{a}$ and $\underline{b}$ the vectors:
   a) $\overrightarrow{BC}$
   b) $\overrightarrow{AD}$

2. Write down the vector $\overrightarrow{FE}$.

3. What geometrical fact is exhibited by the vectors $\overrightarrow{FE}$ and $\overrightarrow{AD}$?

AVERAGES

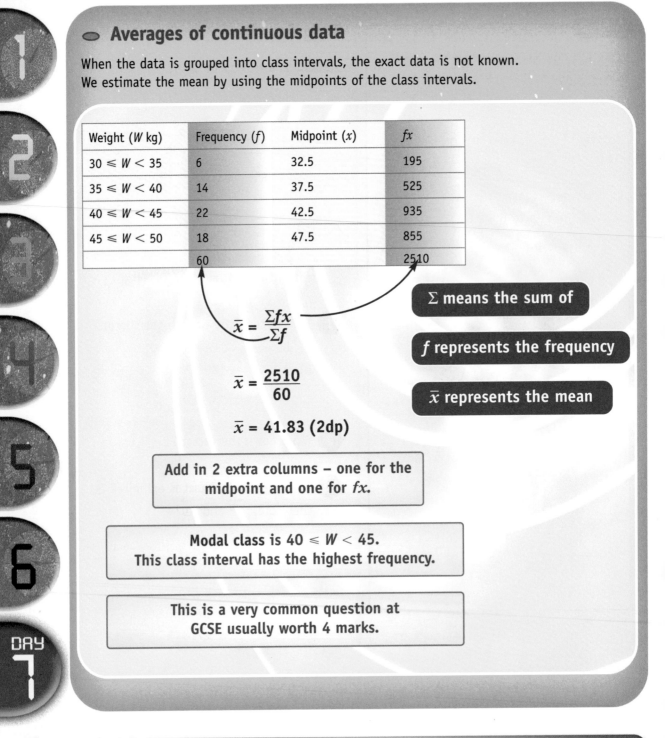

Averages of continuous data

When the data is grouped into class intervals, the exact data is not known. We estimate the mean by using the midpoints of the class intervals.

Weight (W kg)	Frequency (f)	Midpoint (x)	fx
$30 \leqslant W < 35$	6	32.5	195
$35 \leqslant W < 40$	14	37.5	525
$40 \leqslant W < 45$	22	42.5	935
$45 \leqslant W < 50$	18	47.5	855
	60		2510

$$\bar{x} = \frac{\Sigma fx}{\Sigma f}$$

Σ means the sum of

f represents the frequency

$\bar{x}$ represents the mean

$$\bar{x} = \frac{2510}{60}$$

$$\bar{x} = 41.83 \text{ (2dp)}$$

Add in 2 extra columns – one for the midpoint and one for *fx*.

Modal class is $40 \leqslant W < 45$.
This class interval has the highest frequency.

This is a very common question at GCSE usually worth 4 marks.

You need to understand the difference between mean, median and mode and be able to find them in the examination.

10 MINS

Moving averages

- Used to smooth out the changes in a set of data that varies over a period of time.
- A four-point moving average uses four data items in each calculation, a three-point moving average uses three and so on.

Example

Find the four-point moving average for the following data:

3 2 0 1 4 6

Average for 1st 4 data points
$(3 + 2 + 0 + 1) \div 4 = 1.5$

Average for data points 2 to 5
$(2 + 0 + 1 + 4) \div 4 = 1.75$

Average for data points 3 to 6
$(0 + 1 + 4 + 6) \div 4 = 2.75$

- Used to show the trend in a set of data.
- Can be used to draw a trend line on a time series graph.

Progress check

1 Complete the statements for this set of data:

2, 7, 1, 4, 2, 2, 3, 1, 2, 4

a) the mean of the data is

....

b) the mode of the data is

....

c) the range of the data is

....

d) the median of the data is

....

2 The heights, h cm, of some students are shown in the table.

Height (cm)	Frequency
$140 \leq h < 145$	4
$145 \leq h < 150$	9
$150 \leq h < 155$	15
$155 \leq h < 160$	6

Calculate an estimate for the mean of this data.

3 Work out the three-point moving averages for these data:

2, 1, 3, 4, 5, 6

DAY 7

CUMULATIVE FREQUENCY GRAPHS AND BOX PLOTS 1

Example

The table shows the marks of 94 students in a Mathematics exam.

a) Complete the cumulative frequency table for this data.

Mark	Frequency	Mark	Cumulative Frequency	
0–20	2	⩽ 20	2	2
21–30	6	⩽ 30	8	(2 + 6)
31–40	10	⩽ 40	18	(2 + 6 + 10)
41–50	17	⩽ 50	35	(2 + 6 + 10 + 17)
51–60	24	⩽ 60	59	(2 + 6 + 10 + 17 + 24)
61–70	17	⩽ 70	76	(2 + 6 + 10 + 17 + 24 + 17)
71–80	11	⩽ 80	87	(2 + 6 + 10 + 17 + 24 + 17 + 11)
81–90	4	⩽ 90	91	(2 + 6 + 10 + 17 + 24 + 17 + 11 + 4)
91–100	3	⩽ 100	94	(2 + 6 + 10 + 17 + 24 + 17 + 11 + 4 + 3)

This means that 87 students had a score of 80 or less

Cumulative frequency is a running total of all the frequencies.

b) Draw a cumulative frequency graph for this data.

- To do this we must plot the top value of each class interval on the x-axis and the cumulative frequency on the y-axis.

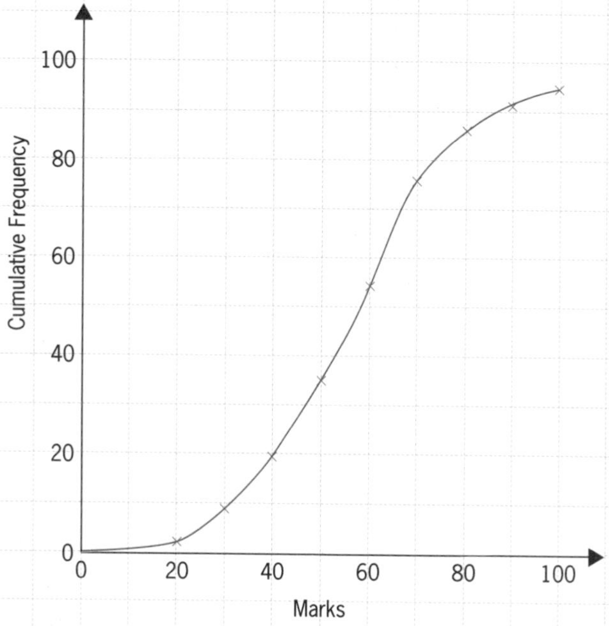

- Plot (20, 2) (30, 8) (40, 18) ...

- Join the points with a smooth curve.

- Since no students had less than zero marks, the graph starts at (0, 0).

CUMULATIVE FREQUENCY GRAPHS AND BOX PLOTS 2

With a cumulative frequency graph it is possible to estimate the median of grouped data and the interquartile range.

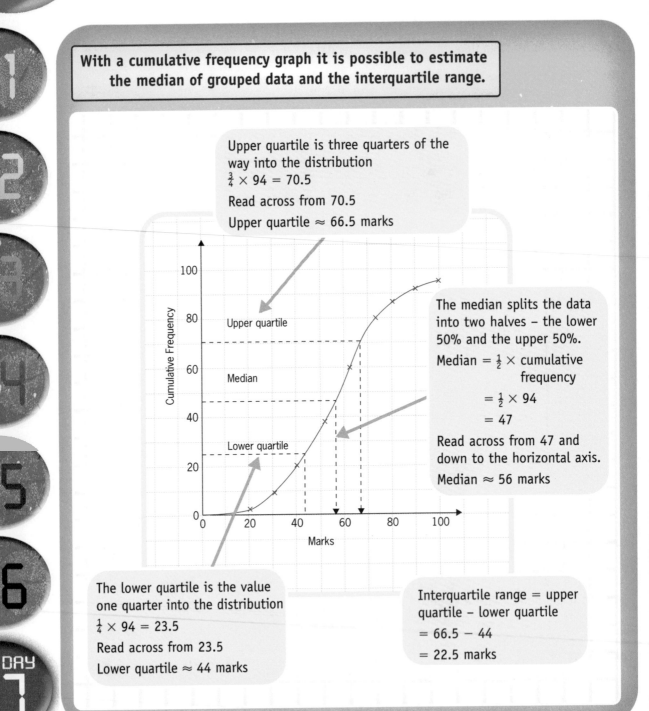

Upper quartile is three quarters of the way into the distribution

$\frac{3}{4} \times 94 = 70.5$

Read across from 70.5

Upper quartile $\approx$ 66.5 marks

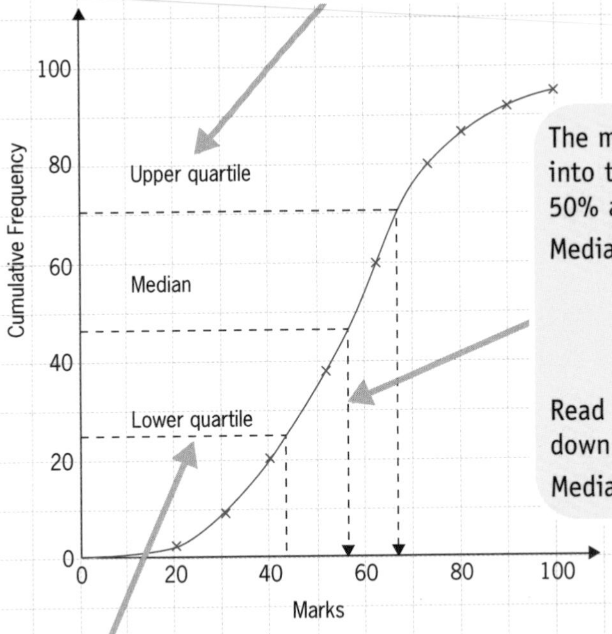

The median splits the data into two halves – the lower 50% and the upper 50%.

Median $= \frac{1}{2} \times$ cumulative frequency

$= \frac{1}{2} \times 94$

$= 47$

Read across from 47 and down to the horizontal axis.

Median $\approx$ 56 marks

The lower quartile is the value one quarter into the distribution

$\frac{1}{4} \times 94 = 23.5$

Read across from 23.5

Lower quartile $\approx$ 44 marks

Interquartile range = upper quartile – lower quartile

$= 66.5 - 44$

$= 22.5$ marks

- A large interquartile range indicates that the 'middle half' of the data is widely spread about the median.
- A small interquartile range indicates that the 'middle half' of the data is concentrated about the median.

Box plots

- Sometimes known as box and whisker diagrams.
- Cumulative frequency graphs are not easy to compare; a box plot shows the Interquartile range as a box.

Examples

The box plot of the above cumulative frequency graph would look like this:

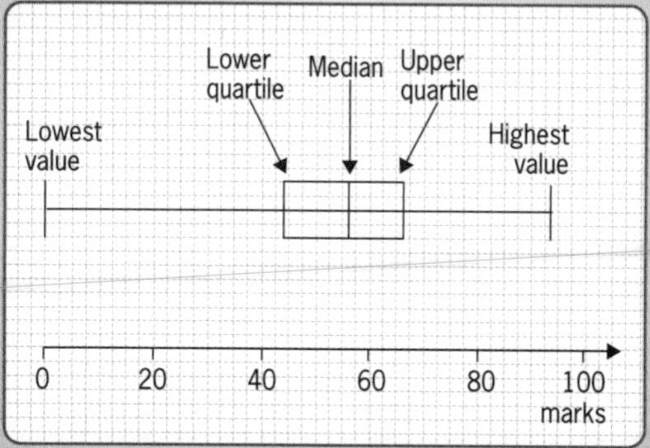

Progress check

The times in minutes to finish an assault course are listed in order

8, 12, 12, 13, 15, 17,
22, 23, 23, 27, 29

1. Find these:
   a) the lower quartile
   b) the interquartile range

2. Draw a box plot for these data:

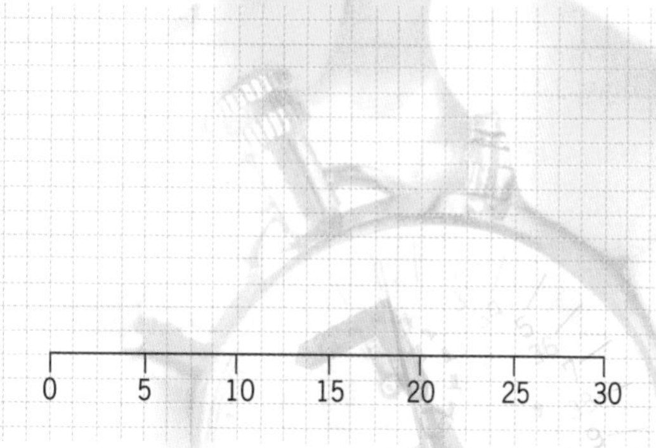

HISTOGRAMS

Drawing histograms

- When the widths of the bars are different the vertical axis is known as the frequency density.

$$\text{frequency density} = \frac{\text{frequency}}{\text{class width}}$$

- The areas of the rectangles are equal to the frequencies they represent.

Example

The table shows the time in seconds it takes people to swim 100 metres. Draw a histogram of this information.

Time, t (seconds)	Frequency	Frequency density
$100 < t \leqslant 110$	2	$2 \div 10 = 0.2$
$110 < t \leqslant 140$	24	$24 \div 30 = 0.8$
$140 < t \leqslant 160$	42	$42 \div 20 = 2.1$
$160 < t \leqslant 200$	50	$50 \div 40 = 1.25$
$200 < t \leqslant 220$	24	$24 \div 20 = 1.2$
$220 < t \leqslant 300$	20	$20 \div 80 = 0.25$

- To draw a histogram you need to calculate the frequency densities. Add an extra column to the table.

Histograms are similar to bar charts except their bars can be different widths. The area of each 'bar' represents the frequency.

15 MINS

- Draw on graph paper – make sure there are no gaps!

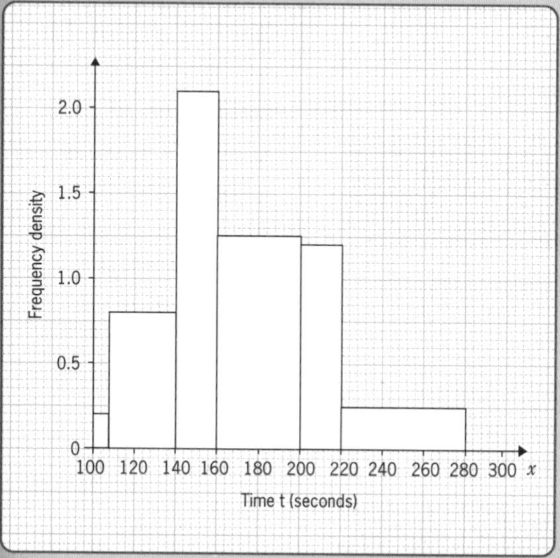

- Sometimes you will be asked to read from a histogram. In this case rearrange the formula for frequency density.

frequency = frequency density x class width

Progress check

The table and histogram give information about the distance (d km) travelled to work by some employees.

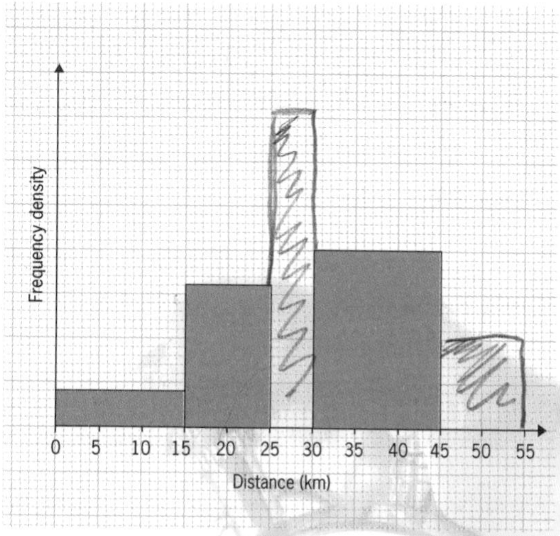

Distance (km)	Frequency	Frequency density
$0 < d \leqslant 15$	12	0.8
$15 < d \leqslant 25$	....	
$25 < d \leqslant 30$	36	7.2
$30 < d \leqslant 45$	....	
$45 < d \leqslant 55$	20	2

1. Use the information in the histogram to complete the table.

2. Use the table to complete the histogram.

1 2 3 4 5 6 DAY 7

PROBABILITY

Tree diagrams

Tree diagrams are used to show the possible outcomes of two or more events. There are two rules you need to know first.

The OR rule

If two or more events are mutually exclusive the probability of A or B happening is found by adding the probabilities.

P (A or B) = P (A) + P (B)
(This also works for more than two outcomes)

The AND rule

If two or more events are independent, the probability of A and B and C happening together is found by multiplying the separate probabilities

P (A and B and C ...) = P (A) x P (B) x P (C) ...

Example

A bag contains 3 red and 4 blue counters. A counter is taken from the bag at random, its colour is noted and then it is replaced in the bag. A second counter is then taken out of the bag. Draw a tree diagram to illustrate this information.

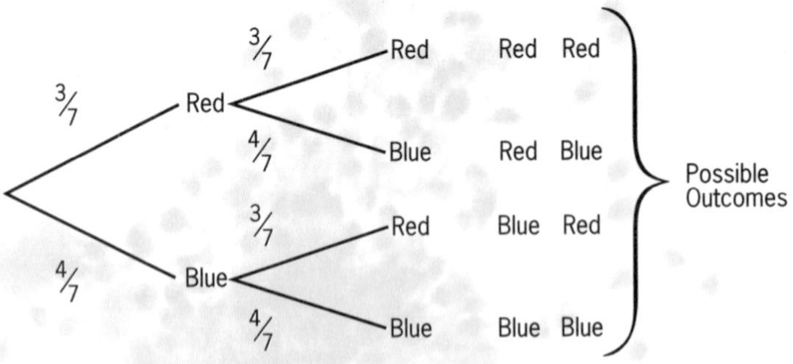

Remember that the branches leaving each point on the tree add up to 1.

Probability is the chance or likelihood that something will happen. All probabilities lie between 0 and 1.

10 MINS

Work out the probability of:

i) picking two blues

ii) picking one of either colour

i) P(two blues) = P(B) × P(B)

$$= \frac{4}{7} \times \frac{4}{7}$$

$$= \frac{16}{49}$$

Remember to multiply along the branches.

ii) P(one of either colour)

$$= P(B) \times P(R)$$

$$= \frac{4}{7} \times \frac{3}{7}$$

$$= \frac{12}{49}$$

OR

OR means add

P(R) × P(B)

$$= \frac{3}{7} \times \frac{4}{7}$$

$$= \frac{12}{49}$$

$$= \frac{12}{49} + \frac{12}{49}$$

P (one of either colour)

$$= \frac{24}{49}$$

Remember to include all possibilities.

Progress check

Charlotte has a biased die. The probability of getting a three is 0.4. She rolls the die twice.

1 Complete the tree diagram.

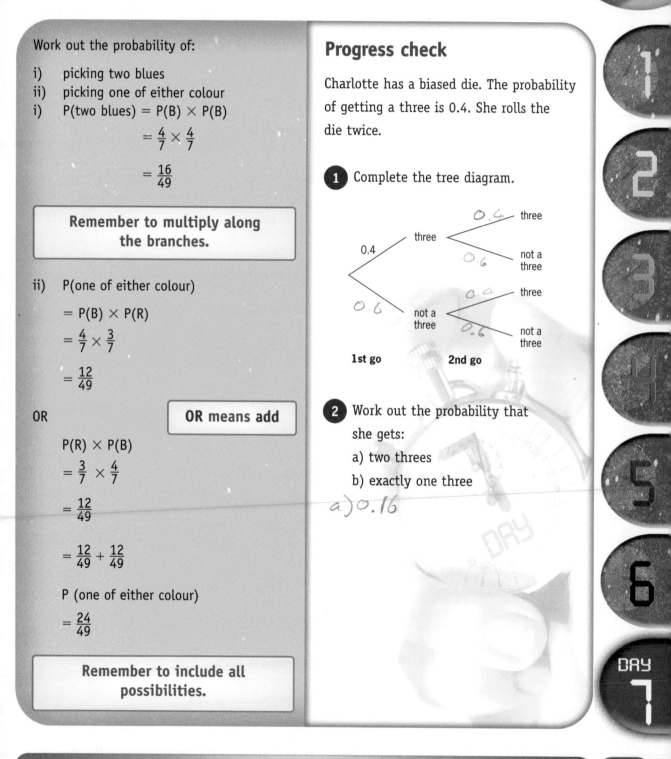

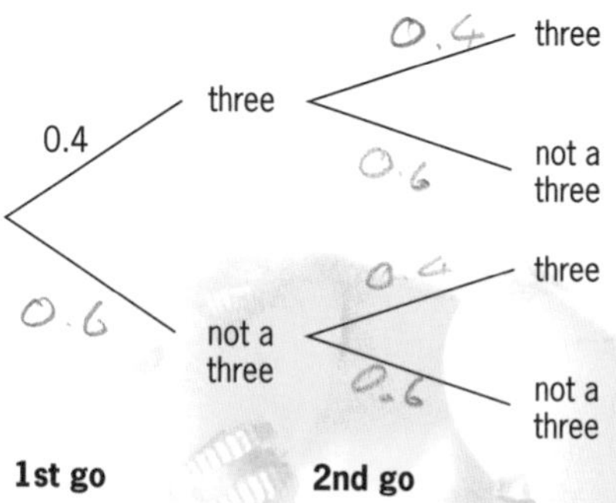

2 Work out the probability that she gets:

a) two threes

b) exactly one three

a) 0.16

1 2 3 4 5 6 DAY 7

ANSWERS

Prime factors, HCF and LCM

1 a) $50 = 2 \times 5^2$
   b) $360 = 2^3 \times 3^2 \times 5$
   c) $16 = 2^4$

2 a) False
   b) True
   c) True
   d) False

3 HCF = 12 LCM = 144

Fractions and recurring decimals

1 a) $\frac{13}{15}$ b) $\frac{11}{21}$
   c) $\frac{10}{63}$ d) $\frac{81}{242}$

2 a) $\frac{7}{9}$ b) $\frac{215}{999}$ c) $\frac{16}{45}$

Repeated percentage change and compound interest

1 £5518.28

2 $(1.027)^2 = 1.054729$

3 £121 856

Reverse percentage problems

1 a) £52.77
   b) £106.38
   c) £208.51
   d) £446.81

2 a) correct
   b) incorrect
   c) correct

Indices

1 a) 6^8 b) 12^{13} c) 7^{24} d) 5^6 e) 2^5

2 a) $6b^{10}$ b) $2b^{-16}$ c) $9b^8$ d) $6b^{14}$
   e) $\frac{1}{25x^4y^6}$

Standard index form

1 a) 6.4×10^4
   b) 2.71×10^5
   c) 4.6×10^{-4}
   d) 7.4×10^{-8}

2 a) 1.2×10^{11}
   b) 2×10^{-1}
   c) 3.5×10^{16}

3 a) 1.4375×10^{18}
   b) 5.48×10^{19}

Surds

1 a) $2\sqrt{6}$
   b) $10\sqrt{2}$
   c) $6\sqrt{3}$

2 Molly is incorrect since $(2 - \sqrt{3})^2 = (2 - \sqrt{3})(2 - \sqrt{3}) = 4 - 4\sqrt{3} + 3 = 7 - 4\sqrt{3}$

3 $-\frac{8}{3}$

4 Correct since $\frac{1}{\sqrt{2}}$ has been rationalised to give $\frac{\sqrt{2}}{2}$

Proportionality

1 a) $y = kx$
   b) $y = \frac{k}{\sqrt[3]{x}}$
   c) $y = \frac{k}{x}$
   d) $y = kx^3$

2 $a = k\sqrt{x}$
   $8 = k\sqrt{4}$
   $\therefore k = 4$
   $a = \pm 4\sqrt{x}$
   $64 = 4\sqrt{x}$
   $x = 256$

Upper and lower bounds of measurement

1 A – upper bound
  E – lower bound
2 Upper bound = 0.226
  Lower bound = $0.2\dot{1}\dot{8}$

Formulae

1 a) $\frac{-31}{5} = -6\frac{1}{5}$
  b) 4.36
  c) 9
2 $u = \pm\sqrt{v^2 - 2as}$
3 $p = \frac{-(t + vq)}{q - 1}$ or $\frac{(t + vq)}{1 - q}$

Brackets and factorisation

1 a) $x^2 + x - 6$
  b) $6x - 8$
  c) $4x^2 - 12x$
  d) $x^2 - 6x + 9$
2 a) $4x(x + 2)$
  b) $6x(2y - x)$
  c) $3ab(a + 2b)$
3 a) $(x + 2)(x + 2)$
  b) $(x - 2)(x - 3)$
  c) $(x + 1)(x - 5)$

Equations

1 $x = 8$
2 $x = -5$
3 $x = -\frac{1}{2}$
4 $x = -3.25$
5 $x = -\frac{1}{2}$
6 $x = 17$
7 $x = 5.5$ cm, shortest length:
  $2x - 5 = 6$ cm
8 $k + 3 = \frac{1}{3}$
  $k = -\frac{8}{3}$
9 $4^{2k} = 4^3$
  $2k = 3$
  $k = \frac{3}{2}$
10 $2^{3k-1} = 2^6$
  $3k - 1 = 6$
  $k = \frac{7}{3}$

Solving quadratic and cubic equations

1 3.3 (1 dp)
2 a) $x = -5$, $x = 3$
  b) $x = -\frac{1}{2}$, $x = -2$
3 a) $(x + 4)^2 - 14$ $\therefore a = 4, b = -14$
  b) −14

Simultaneous linear equations

1 a) $b = -4.5$ $a = 4$
  b) $p = -3$, $r = 4$
  c) $x = 2$, $y = 3$
2 $x = 2$, $y = 4$

Solivng linear and quadratic equations simultaneously

1 a) $x = -1$, $y = 3$
    $x = 2$, $y = 6$
  b) $x = 3$, $y = 4$
    $x = -4$, $y = -3$
2 a) This is where the quadratic graph $y = x^2 + 2$ intersects the straight line graph $y = x + 4$. Their points of intersection are (–1, 3) and (2, 6).
  b) This is where the circle $x^2 + y^2 = 25$ intersects with the straight line $y = x + 1$. Their points of intersection are (3, 4) and (–4, –3).

Algebraic fractions

1 a) No b) yes c) yes

2. a) $\dfrac{5x - 1}{(x + 1)(x - 1)}$

b) $\dfrac{2aq}{3c}$

c) $\dfrac{40(a + b)}{3}$

Inequalities

1 a) $x < 2.2$

b) $\frac{4}{3} \leqslant x < 3$

c) $x > -\dfrac{9}{5}$

2

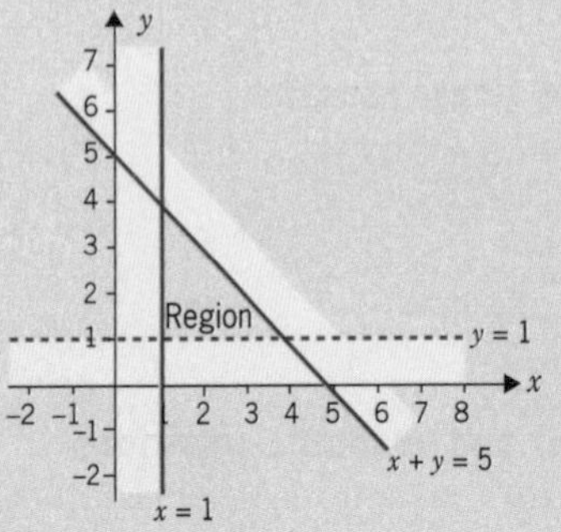

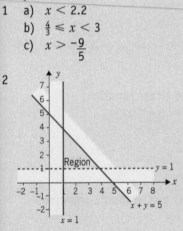

Curved graphs

1 a)

x	−3	−2	−1	0	1	2	3
y	−28	−9	−2	−1	0	7	26

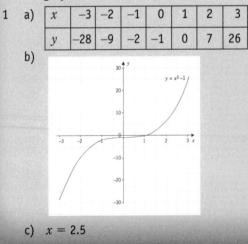

b)

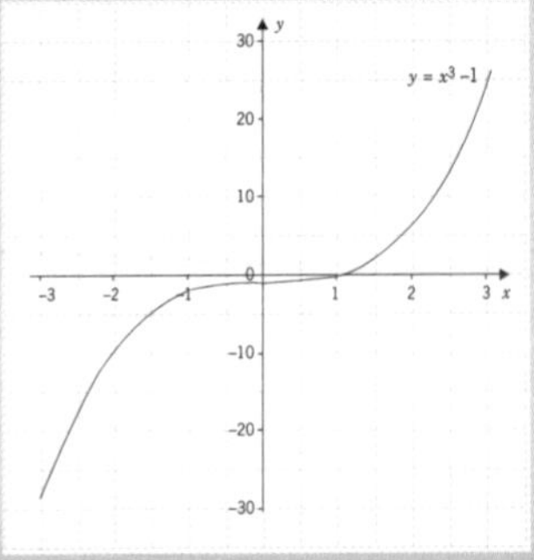

c) $x = 2.5$

2 Graph A: $y = \dfrac{3}{x}$

Graph B: $y = 4x + 2$

Graph C: $y = x^3 - 5$

Graph D: $y = 2 - x^2$

Interpreting graphs

1 True for both solutions

2 False for both solutions

3 True for $x = -0.7$, false for $x = 5.6$

Functions and transformations

1 (−1, −3)

2 (2, −7)

3 (4, −3)

4 (−2, −3)

5 (1, −3)

Loci

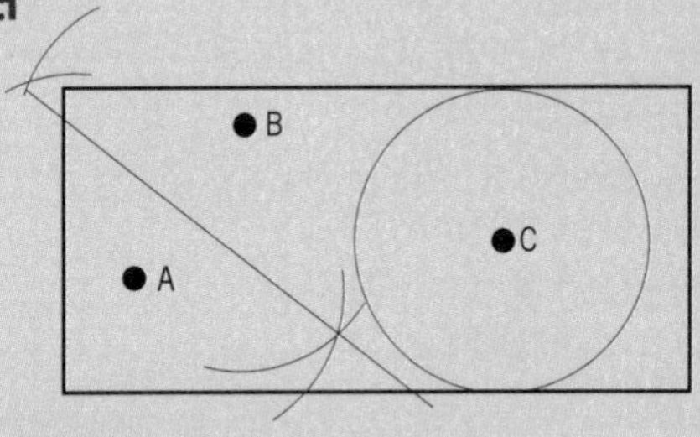

Enlargements

1

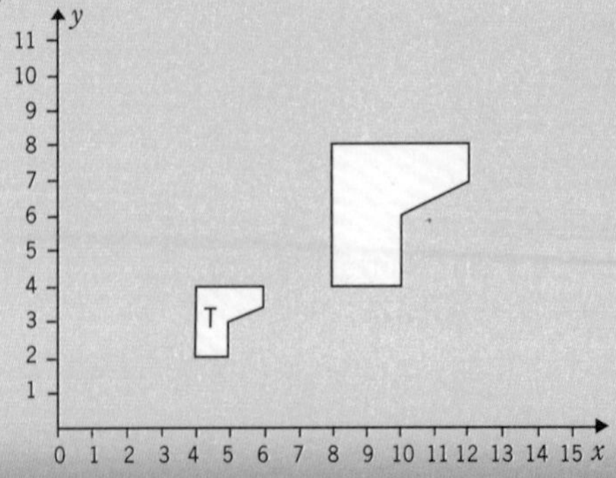

2

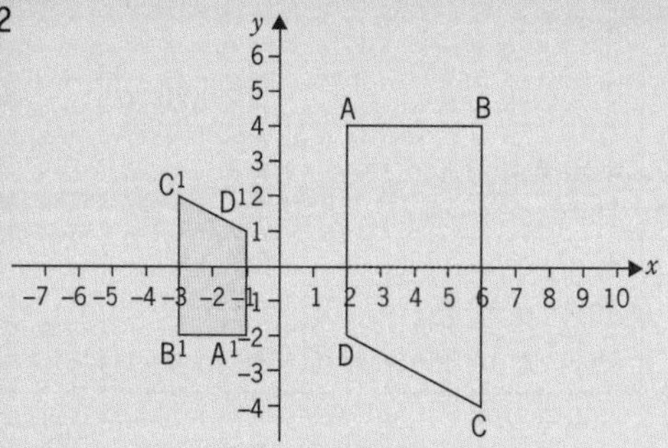

Similarity and congruency

1 a) 20.9 cm
   b) 13.8 cm
   c) 4.5 cm

2 28.4 cm^2 (3sf)

3 Yes congruent : SAS i.e. two sides and the included angle.

Circle theorems

a) 62° b) 109° c) 53° d) 50° e) 126°

Pythagoras' theorem

1 Since $26^2 = 24^2 + 10^2$
   $676 = 576 + 100$ and this obeys Pythagoras' theorem, then the triangle must be right-angled.

2 15.3 cm

3 $\sqrt{149}$

Trigonemtry in 3-D figures

1 a) True
   b) False
   c) False
   d) True

2 10° (nearest degree)

The size and cosine rules

1 a) Correct
   b) Correct
   c) Incorrect

2 a) 114° (nearest degree)
   b) 53° (nearest degree)

Trigonometric functions

1

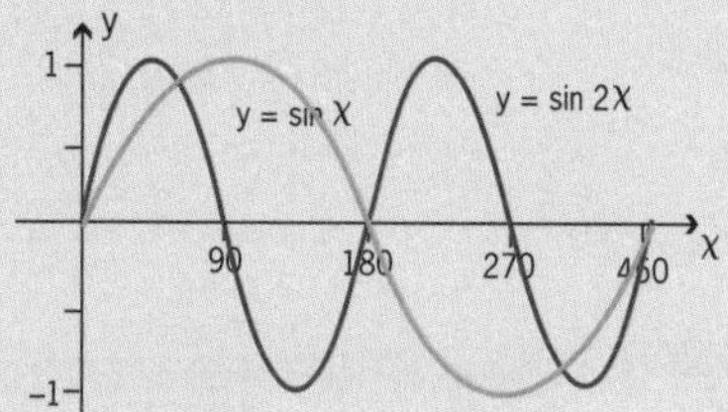

2 60°, 300°

3 –225°, 45°, 135°

Arc, sector and segment

1 a) 14.05 cm
   b) 49.17 cm^2
   c) 26.97 cm^2

2 True since

$$\left(\frac{70°}{360°}\times\pi\times8^2-\frac{1}{2}\times8\times8\times\sin 70°\right)$$

$= 9.03\ cm^2$

Surface area and volume

1 565 cm^3

2 637 cm^3

3 905 cm^3

4 245 cm^3

Dimensions and converting units

1 r^3 represents L^3, which is volume not surface area.

2 a) True
b) False
c) True

3 20 000 cm^2 (3sf)

4 0.6 m^3

Vectors

1 a) $-\underline{a} + \underline{b}$
b) $-2\underline{a} + 2\underline{b}$

2 $-\underline{a} + \underline{b}$

3 Since $\overrightarrow{FE} = -\underline{a} + \underline{b}$
and $\overrightarrow{AD} = 2(-\underline{a} + \underline{b}) = 2\,\overrightarrow{FE}$

Vector $\overrightarrow{AD}$ is twice the length and parallel to $\overrightarrow{FE}$.

Averages

1 a) 2.8
b) 2
c) 6
d) 2

2 150.88

3 2, $2.\dot{6}$, 4, 5

Cumulative frequency graphs and box plots

1 a) lower quartile = 12
b) interquartile range = 11

2

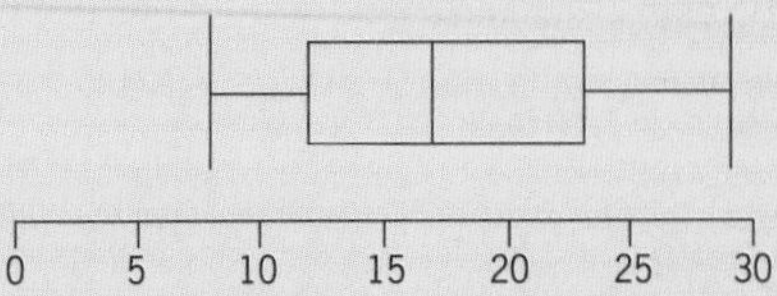

Histograms

1

Distance (km)	Frequency
$0 < d \leqslant 15$	12
$15 < d \leqslant 25$	32
$25 < d \leqslant 30$	36
$30 < d \leqslant 45$	60
$45 < d \leqslant 55$	20

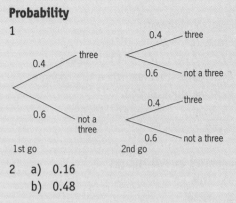

2

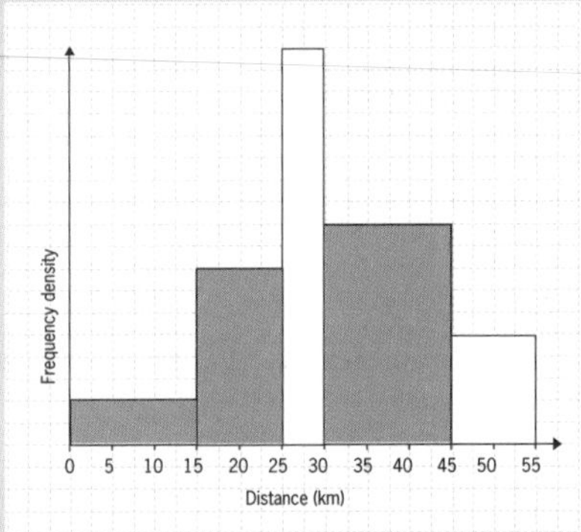

Probability

1

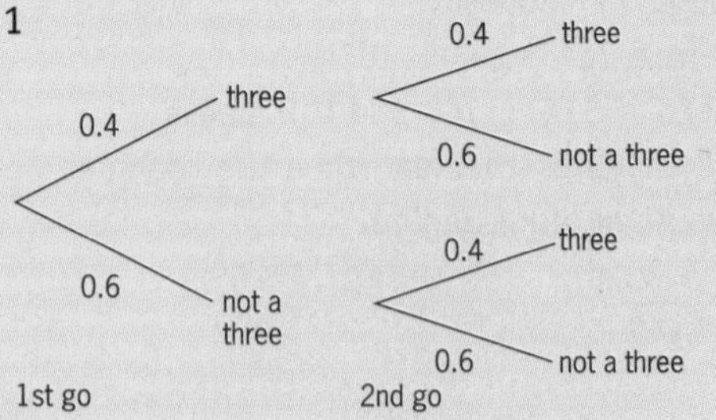

2 a) 0.16
b) 0.48

NOTES

NOTES